RADIOFREQUENCY SPECTRUM MANAGEMENT: BACKGROUND, STATUS, AND CURRENT ISSUES

RADIOFREQUENCY SPECTRUM MANAGEMENT: BACKGROUND, STATUS, AND CURRENT ISSUES

LENNARD G. KRUGER

Novinka Books
New York

Senior Editors: Susan Boriotti and Donna Dennis
Coordinating Editor: Tatiana Shohov
Office Manager: Annette Hellinger
Graphics: Wanda Serrano
Editorial Production: Jennifer Vogt, Ronald Doda, Matthew Kozlowski,
Jonathan Rose and Maya Columbus
Circulation: Ave Maria Gonzalez, Vera Popovich, Luis Aviles, Melissa Diaz,
Vladimir Klestov and Jeannie Pappas
Communications and Acquisitions: Serge P. Shohov

Library of Congress Cataloging-in-Publication Data
Available Upon Request

ISBN: 1-59033-353-5.

Copyright © 2002 by Novinka Books, An Imprint of
Nova Science Publishers, Inc.
400 Oser Ave, Suite 1600
Hauppauge, New York 11788-3619
Tele. 631-231-7269   Fax 631-231-8175
e-mail: Novascience@earthlink.net
Web Site: http://www.novapublishers.com

CONTENTS

PREFACE

The radio spectrum, a limited and valuable resource, is used for all forms of wireless communications including cellular telephony, paging, personal communications service, radio and television broadcast, telephone radio relay, aeronautical and maritime radio navigation, and satellite command and control.

The federal government manages the spectrum to maximize efficiency in its use and to prevent interference among spectrum users. The National Telecommunications and Information Administration (NTIA) manages all spectrum used by the federal government and the Federal Communications Commission (FCC) manages all non-federal spectrum (used by state and local governments and the commercial sector).

For several years, the FCC has been using auctions to assign certain commercial spectrum licenses, instead of providing the licenses for free, raising over $31 billion in actual and expected cash receipts. Although some auctions have been criticized, by most assessments, auctions are considered more effective than previous methods, both in terms of the speed with which the licenses are distributed and the revenue that can be raised. As radio technology improves, higher frequencies may become available, and spectrum may be utilized more efficiently, contributing to the already increasing demand for wireless services.

INTRODUCTION

The radiofrequency spectrum, a limited and valuable resource, is used for all forms of wireless communication, including cellular telephony, radio, and television broadcast, telephone radio relay, aeronautical and maritime radio navigation, and satellite command, control, and communications. The radiofrequency spectrum (or simply, the "spectrum") is used to support a wide variety of applications in commerce, federal, state, and local government, and interpersonal communications. Because the spectrum cannot support all of these uses simultaneously to an unlimited extent, its use must be managed to prevent signal interference. The growth of telecommunications and information technologies and services has led to an ever increasing demand for the use of spectrum among competing businesses, government agencies, and other groups. As a result, the spectrum, which is regulated by the federal government, has become increasingly valuable. The need for managing the spectrum, including its allocation, has received growing attention by Congress in recent years.

Chapter 1

SPECTRUM TECHNOLOGY BASICS

Electromagnetic radiation is the propagation of energy that travels through space in the form of waves. The most familiar form is light, called the visible spectrum. The **radiofrequency** spectrum is the portion of electromagnetic spectrum that carries radio waves. Figure 1 shows the radio spectrum as part of the measured electromagnetic spectrum. **Wavelength** is the distance a wave takes to complete one cycle. **Frequency** is the number of waves traveling by a given point per unit of time, in cycles per second, or **hertz** (Hz).[1] The relationship between frequency (f) and wavelength (λ) is depicted in Figure 2. **Bandwidth** is a measure of how fast data is transmitted or received whether through wires, air or space. Signals are transmitted over a range of frequencies which determines the bandwidth of the signal. Thus a system that operates on frequencies between 150 and 200 MHZ has a bandwidth of 50 MHZ.[2] In general, the greater the bandwidth, the more information that can be transmitted.

---

[1] Radiofrequency is usually measured in kilohertz (kHz), which is thousands of hertz, megahertz (MHz) which is millions of hertz, and gigahertz (GHz) which is billions of hertz.

[2] Bandwidth is also measured in bits per second (bps) instead of cycles per second, especially in digital systems.

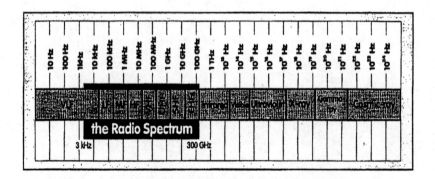

Figure 1. The Electromagnetic Spectrum

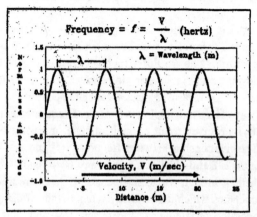

Figure 2. Frequency vs. Wavelength

An important distinction in spectrum technology is the difference between narrowband and broadband. **Narrowband** signals have a smaller bandwidth (on the order of kHz) and are used for limited services such as paging and low-speed data transmission. **Broadband** signals have a large bandwidth (on the order of MHZ) and can support many advanced telecommunications services such as high-speed data and video transmission. The precise dividing line between broadband and narrowband is not always clear, and changes as technology evolves.

Two other important terms are analog signals and digital signals, depicted in Figure 3. In **analog** signal transmissions, information (sound, video, or data) travels in a continuous wave whose strength and frequency vary directly with a changing physical quantity at the source (i.e., the signal is directly analogous to the source). In **digital** signals, information is converted to ones and zeros which are formatted and sent as electrical impulses. Advantages of using digital signals include greater accuracy, reduction in noise (unwanted signals) and a greater capacity for sending information. Analog signals have the advantage of greater fidelity to the source, although that advantage can be made very small by increasing the rate at which signals are digitized. Digital signals are acknowledged to be superior to analog signals for the majority of applications.[3]

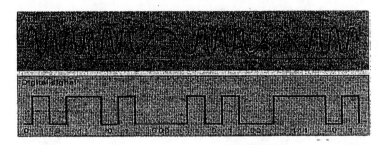

Figure 3. Schematic Comparing Analog vs. Digital Signals

Electromagnetic waves have many characteristics that govern how spectrum can be used in telecommunications systems. For example, antennas are used for transmitting and receiving signals, and can be designed to transmit in all directions or can be directed toward specific receivers. Receiving antennas are typically aligned with the transmitting antenna to maximize signal reception, but unintended signals can still interfere with the reception of the information sent. To avoid signal **interference** from stray signals, more than one radio signal usually cannot be transmitted in the same frequency range, at the same time, in the same area. Another characteristic is that the spectrum, unlike other natural resources, is not destroyed by use. As soon as one user stops transmitting signals over a portion of the spectrum, another can immediately re-use it. The spectrum is scarce, however, because at any given time and place, one use of a frequency precludes its use for any other purpose.

---

[3] For further discussion see CRS Report 96-401 SPR, Telecommunications Signal Transmission: Analog vs. Digital, May 7, 1996.

Chapter 2

USES OF THE RADIO SPECTRUM

Spectrum is used to provide a variety of wireless communications services, which are categorized as fixed or mobile voice/data services or broadcast services. Demand for all wireless services has grown rapidly in recent years in both terrestrial and satellite applications. Federal agencies use spectrum for various purposes, including military and national security needs, weather radio services, radars and communication systems to control commercial and private air and maritime traffic, weather satellite systems, flood warning and water control systems, and time signals. Examples include communications between defense platforms (e.g., aircraft and ships) and military bases, Voice of America broadcasts of politically oriented information in foreign countries, and data transmissions by the Department of Energy to monitor electrical power grids. Uses of the spectrum by state and local governments, and commercial entities are also varied and pervasive.

Much of the radiofrequency spectrum is shared among two or more wireless services. In these cases, one service may be designated as primary and the other services using the same frequencies as secondary, if the secondary services are prohibited from causing harmful radio interference to the operations of the primary service. For example, bands assigned to federal radar operations are primary, while at the same time, amateur radio spectrum is assigned secondary status within those same bands. Because these two systems do not interfere with each other, both may be designated as co-primary if they are given an equal degree of protection from harmful interference. Harmful interference is defined as any radio signal that degrades, obstructs, or interrupts the service provided by another operation.[4]

COMMERCIAL VOICE AND
DATA TRANSMISSION SERVICES

- *Cellular telephone* systems consist of an array of terrestrial base stations (each covering a geographic area called a *cell*) that transmit and receive signals to and from mobile or fixed wireless telephones to provide two-way voice and data communications over a geographic region. Each cellular system is comprised of a cluster of cells of varying sizes according to the number of users in the area and the local terrain. All of the cells are connected to a mobile telephone switching office that manages all communications traffic in the cellular network and connects to public switched telephone network (the "wireline" network).

- *Paging* is a low-cost one-way message-sending system that uses base stations similar to cellular telephone systems. An enhanced paging system, called *messaging*, has a limited two-way data transmission capability.

- *Personal communications service (PCS)* is a wireless telephone services similar to cellular technology but using higher frequencies (around 1900 MHZ, compared to 800 MHZ for cellular services) and digital signal transmission technology (many cellular services are converting to digital). From the users' perspective, however, there is little difference between PCS and cellular services. Narrowband PCS can provide two-way messaging for interactive low-speed data applications (such as e-mail) but generally not voice, while broadband PCS provides a wider range of services.

- *Interactive video and data services (IVDS, not called 2-8-219 MHZ service)* is a subscription service that allows viewers cable and broadcast television to interact with the transmitting point over short distances. Applications include ordering goods or services

---

[4] Code of Federal Regulations Title 47, Part 2, General Regulations, Sec. 2.1 Terms and

offered by television programming, viewer polling, remote meter reading, vending inventory control, and cable television signal theft deterrence. These services now may also provide Internet access.

- *Specialized mobile radio (SMR)* is a wireless service originally created for public safety and dispatch communications. Newer enhanced SMR systems connect to the public telephone network to compete against cellular and PCS.

- *Satellite systems* provide communications to very large regions using signal transmissions between satellites and ground facilities. Communications satellites are used for voice, data, and broadcast purposes for government and commercial operations. Geostationary satellites maintain a fixed position relative to a point on Earth, and provide communications services for users at fixed or mobile locations. Low and medium Earth orbiting satellite systems are also being developed to provide fixed or mobile communications services including paging, voice, fax, and interactive Internet services.

Additionally, the Federal Communications Commission (FCC) has made spectrum available for unlicensed data services used for low-power applications. Providing this free spectrum for unlicensed data services is known to stimulate entrepreneurial activity, and use of this spectrum is intensive. Devices allowed to operate without a license[5] include cordless telephones, hearing aids, citizens band radio, consumer digital devices, industrial, scientific, and medical equipment, family radio service, and other innovations. Unlicensed spectrum is also used for wireless computing, whereby portable laptop computers interact with mainframes or wireless local area networks (LANs). In 1998, the FCC made an additional 300 MHZ in the 5 GHz band available for Unlicensed National Information Infrastructure (NII) devices to facilitate wireless access to the Internet and stimulate the development of new devices. Recently, the FCC has modified its rules to permit a greater use of spread spectrum devices (i.e., devices that operate by spreading the communications signal over a wide range of frequencies, thereby reducing the possibility of creating interference with other wireless systems). The FCC has also made extremely high frequency

Definitions.
[5] These are known as Part 15 devices because the rules for the operation of these devices are provided in the Code of Federal Regulations Title 47, Part 15.

spectrum (above 40 GHz) available for future unlicensed applications and has proposed making additional spectrum available above 40 GHz.

COMMERCIAL BROADCAST SERVICES

- *Radio,* the oldest broadcast service, includes the AM (535 to 1,605 kHz) and FM (88 to 108 MHZ) bands within which licenses are assigned at specific frequencies for terrestrial radio broadcast stations. The same frequencies can be used by multiple stations if the stations have sufficient geographic separation. The radio industry is developing new digital audio broadcasting (DAB) technology using the same spectrum bands, and the FCC is developing rules for DAB services.

- *Broadcast television* includes over 1600 currently licensed full service TV stations occupying 402 MHZ in the VHF and UHF bands. Each TV station has a 6 MHZ license. Television broadcasters are now starting to provide new digital television services using the vacant portion of the same band of spectrum.

- *Multipoint distribution service (MDS),* also called wireless cable, is a television broadcast system using digital encrypted signal transmissions in the microwave band (2 to 3 GHz), and providing up to 100 TV channels. MDS includes both single channel and multi-channel MDS *(MMDS)* applications (a more commonly used term). MDS now also offers subscriptions to high speed Internet access and data transmission services. A related service call instructional television fixed service (ITFS) operates under similar technology, but offers more educational programming. In addition, the FCC recently created a new service called Multichannel Video Distribution and Data Service (MVDDS) which uses a technology and system similar to MDS and is permitted to operate on spectrum to be shared with non-geostationary satellite orbit fixed satellite service.[6]

---

[6] FCC Makes Spectrum Available for New Fixed Satellite Service at Ku Band; Seeks Comment on Licensing New Fixed Service at 12 GHz. FCC News Release November 30, 2000.

- *Direct broadcast satellite (DBS)* is a high-powered satellite television delivery system in which consumers receive programming by small (18 inch) receiving antennas in the 12.2-12.7 GHz band. It is related to the older direct-to-home (DTH) satellite television services that use large receiving antennas and operate in several other frequency bands.

- *Digital Audio Radio Services (DARS)* is a new high-fidelity radio service offered by several companies to be delivered to automobiles and other locations by geostationary satellites.

- *Local multipoint distribution service (LMDS)*, also called cellular television, is a new video distribution service for urban areas. Using a cellular architecture, LMDS can also provide two-way telephony (to compete with cellular telephone services), teleconferencing, telemedicine, and data services.

Chapter 3

MANAGEMENT OF THE RADIO SPECTRUM

Since the beginning of the 20th century, when radio broadcast signals were first transmitted, it was realized that the spectrum was a public resource. Governmental entities have assumed responsibility for managing it to avoid interference. Spectrum is managed internationally by the International Telecommunication Union (ITU, a specialized agency of the United Nations located in Geneva, Switzerland). The ITU maintains a Table of Frequency Allocations which identifies spectrum bands for about 40 categories of wireless services with a view to avoiding interference among services. The ITU sponsors biannual World Radio Conferences to update the Table in response to changes in needs and demand for spectrum.

Once the broad categories are established, each country must allocate spectrum for various services within its own borders in compliance with the ITU Table of Frequency Allocations. In the United States, the Communications Act of 1934 established the FCC as an agency independent from the executive brand, to manage all non-federal government spectrum (which includes commercial, state and local government uses), and preserved the President's authority to mange all spectrum used by the federal government. The President also manages frequency assignments to foreign embassies and regulates the characteristics and permissible uses of the government's radio equipment. The President delegates this authority to the Assistance Secretary of Commerce for Communications and Information who is also Administrator of the National Telecommunications and Information Administration (NTIA).

The Communications Act of 1934 directs the FCC to develop classifications for radio services, to allocate frequency bands to various services, and to authorize frequency use. The Act does not, however, mandate specific allocations of bands for federal or non-federal use, which is usually decided through agreements between NTIA and the FCC (although Congress has occasionally directed the transfer of specific spectrum bands from federal to commercial use). The Act authorizes the FCC to grant licenses for radio frequency bands, but provides few details other than requiring that FCC rulings be consistent with the "public interest, convenience, and necessity." The Act authorizes the FCC to regulate "so as to make available...a rapid, efficient, nationwide, and worldwide wire and radio communication service with adequate facilities at reasonable charges, for the purpose of the national defense, and for the purpose of promoting safety of life and property."

The primary FCC offices that develop and implement spectrum policy are the Mass Media Bureau (which regulates all U.S. television and radio stations), the Wireless Telecommunications Bureau (which manages all domestic commercial wireless services except those involving satellite communications), the International Bureau (handling international telecommunications and satellite policies), and the Office of Engineering and Technology (developing spectrum allocations and policy, experimental licensing, spectrum management and analysis, technical standards, and equipment authorization). The FCC develops rules for spectrum use and other telecommunications regulation through lengthy proceedings in accordance with the Administrative Procedures Act. The FCC's Enforcement Bureau monitors the airwaves to ensure that non-federal users are complying with FCC rules, orders and authorizations.

The NTIA offices that focus on spectrum policy include the Office of International Affairs which represents U.S. interests ion international for a, and the Office of Spectrum Management which develops policies and procedures for domestic spectrum use by the federal government. This entails developing long-range plans and war and readiness plans for spectrum use, chairing the Interdepartment Radio Advisory Committee (IRAC),[7] and representing the United States government at International Telecommunications Union Conferences such as the World Radio Conferences (although other federal agencies also participate).

---

[7] The IRAC is composed of representatives of 20 major federal agencies who develop policies for federal spectrum use.

NTIA assigns frequencies and approves the spectrum needs for all federal government systems to support their mandated missions.[8] NTIA strives to improve federal spectrum efficiency by requiring federal users to use commercial services where possible, promoting the use of new spectrum efficient technologies, developing spectrum management plans, and collecting spectrum management fees (pursuant to congressional mandate). Since most spectrum is shared between government and private sector uses, NTIA (in conjunction with the FCC) is working toward increasing private sector access to the shared program. As a provision of the Omnibus Reconciliation Act of 1993 (47 U.S.C. 927), NTIA has begun to reallocate 235 MHZ of spectrum from federal government use to the private sector (195 MHZ of that amount has been reallocated and the remaining spectrum is scheduled for auction by the FCC in 2002).

SPECTRUM AUCTIONS

Because two or more signal transmissions over the same frequency in the same location at the same time could cause interference (a distortion of the signals), the FCC, over many years, has developed and refined a system of exclusive licenses for users of specific frequencies.[9] Traditionally, the FCC granted licenses using a process known as "comparative hearings" and later using lotteries. After years of debate over the idea of using competitive bidding (i.e., auctions) to assign spectrum licenses, the Omnibus Budget Reconciliation Act of 1993 (47 U.S.C. 927) added Section 309(j) to the Communications Act, authorizing the FCC to use auctions to award spectrum licenses for certain wireless communications services (later expanded by the Balanced Budget Act of 1997). The main category of services for which licenses may be auctioned are called commercial mobile radio services (CMRS) which include PCS, cellular, and most SMR and mobile satellite services. CMRS providers are regulated as common carriers (with some exceptions) to ensure regulatory parity among similar services that will compete against one another for subscribers.[10] The FCC has the

---

[8] Major federal spectrum users include the Departments of Defense, Justice, Transportation, Interior, Agriculture, Commerce, Treasury, Energy, the National Aeronautics and Space Administration, and the Federal Emergency Management Agency.

[9] Technically, two signals will interfere with each other even if they are not at the same exact frequency, but are close in frequency. To avoid harmful interference, the frequencies must have frequencies that are sufficiently different, known as a "minimum separation."

[10] Other services, classified as Private Mobile Radio Services (PMRS), are prohibited from connecting to the public switched telephone network.

authority to conduct auctions only when applications are mutually exclusive (i.e., two licensees in the same frequency band would be unable to operate without causing interference with each other) and services are primarily subscription-based.[11] The FCC does not have authority to conduct auctions for licenses that have already been issued.

Auction Rules

The FCC initially developed rules for each auction separately (with some common elements), but after several years of trial and error, it developed a set of general auction rules and procedures. While there may be special requirements for specific auctions, the following rules generally apply. As a screening mechanism, all auctions require bidders to submit applications and up-front payments prior to the auction. Most auctions are conducted in simultaneous multiple-round bidding, in which the FCC accepts bids on a large set of related licenses simultaneously using electronic communications. Bidders can bid in consecutive rounds on any license offered until all bidding has stopped on all licenses. Even though licenses must be renewed periodically, it is generally understood that license winners will be able to keep the license perpetually, as long as they comply with FCC rules.[12]

For some auctions, the FCC gives special bidding credits to smaller companies, called entrepreneurs, defined as having annual gross revenues of less than $125 million and total assets of less than $500 million. In the first year or so of auctions, the FCC originally also gave special provisions to women-owned, minority-owned, and rural telephone companies (called *designated entities*). After a 1995 Supreme Court decision determined that government affirmative action policies must pass a "strict scrutiny" test to demonstrate past discrimination, the FCC removed those other groups from its list of businesses qualifying for bidding credits.[13] Nevertheless, concerns have been raised that some of the small businesses participating in auctions actually represent larger companies that are excluded from the bidding credits.

---

[11] Licenses are issued for the use of bands of spectrum. In general, a greater bandwidth can carry more information than a smaller bandwidth.

[12] The FCC provides additional information on auctions on its website at http://www.fcc.gov. wtb.auctions.htm.

[13] Adarand Constructors Inc., petitioner v. Federico Pena, Secretary of Transportation, et al. Docket No. 93-1841, decided June 1995.

Service Rules

The FCC also develops services rules for each new service for which a license will be used. Licenses are granted according to the amount of spectrum and the geographic area of coverage. The FCC's plan for the amount of spectrum per license, the number of licenses, and the conditions for use of the designated spectrum, known as the "band plan," is developed for each new wireless service. Licenses can cover small areas, large regions, or the entire nation. Terms used for coverage areas include basic trading areas (BTAs) which correspond roughly to metropolitan areas; major trading areas (MTAs), which are combinations of BTAs dividing the United States into 51 geographic regions of similar levels of commercial activity; and regions, which are combinations of MTAs. Metropolitan statistical areas (MSAs), rural service areas (RSAs), economic areas (EAs), and major economic areas (MEAs) developed by the Department of Commerce for economic forecasts are also used by the FCC to define areas of coverage for some spectrum auctions.

The FCC has also modified some wireless service rules to help new spectrum licensees maximize the value from their licenses. Changes include allowing licensees to partition licenses for greater efficiency, sharing regions among licensees, and expediting the relocation of incumbent microwave licensees from the spectrum purchased in the PCS auctions. The FCC maintains a website on its auction activities at [http://www.fcc.gov/wtb/auctions]. This site provides archived information on all of its completed auctions, details on its ongoing and future auctions, auction-related maps, charts, and service, and auction rules.

COMPLICATIONS WITH SOME AUCTIONS[14]

Despite their general success, the FCC's auctions have experienced several problems from which the FCC has learned and modified subsequent auctions.

---

[14] See archived CRS Report 97-218, Radiofrequency Spectrum Management, updated April 23, 1998, for further analysis of these auctions and other spectrum management issues from a historical perspective.

WCS Auction

By 1997, the FCC had raised over $22 billion from auctions, and many observers in government and the private sector claimed auctions to be a success. That enthusiasm decreased somewhat, however, after the results of two auctions, held in April 1997, for wireless communications services (WCS) and digital audio radio service (DARS). The WCS auction was mandated by the FY1997 Omnibus Appropriations Act (P.L. 104-208 Title III), which directed the FCC to reallocate the use of 30 MHZ of spectrum (some of which had already been allocated for DARS) and to begin the auction for those licenses by April 15, 1997. To implement the Act, the FCC divided the spectrum remaining for DARS in half, surrounded by the newly allocated WCS spectrum. To prevent interference between the two new services, the FCC placed restrictions on the power that could be radiated by WCS. Although the FCC completed the DARS and WCS auctions in April 1997, meeting the congressionally mandated schedule, the revenue obtained was far lower than estimates had previously predicted. Reasons cited for the shortfall included the shortened timetable set by Congress for the FCC to complete the auction, and the technical constraints placed on the WCS spectrum which prevented interference with DARS, but reduced the usefulness of the WCS spectrum.

LMDS Auction and Satellite Spectrum Allocations

The auction for Local Multipoint Distribution Service (LMDS), a new television distribution service for urban areas that may also be used for two-way communications, was held in March 1998. LMDS uses much higher frequencies than existing commercial wireless services (in the 28 and 31 GHz bands). LMDS could compete with cable TV, broadcast TV, MMDS, satellite TV, mobile telephone data, or broadband Internet access services. Finding spectrum for LMDS was complicated due to the spectrum needs of new satellite services in the same bands. At the time when FCC was developing plans for LMDS spectrum, several companies developing new fixed satellite services (FSS) requested the same spectrum for sending signals to their satellites. FSS systems use geostationary and non-geostationary satellites to provide worldwide voice, video, and interactive data services to users at fixed locations. Mobile satellite services (MSS), which serve mobile users as well as fixed users, also wanted these frequencies to interconnect MSS systems to other communications networks, and the FCC had already granted licenses to several of these companies. The FCC decided not to auction the spectrum allocated for FSS or MSS, but divided the 28 GHz band among LMDS, geostationary FSS, non-geostationary FSS, and MSS, in the auction rules for LMDS licenses.[15] A frequency band around 18 GHz was designated for FSS downlink signals to share with several other services. The first LMDS auction raised $577 million; lower than expected by many in the private sector. Because some LMDS licenses received no bids, and other LMDS license winners defaulted on their payments, in 1999 the FCC conducted a re-auction of the remaining and reclaimed licenses, raising an additional $45 million. The reasons for the lower than expected proceeds are not clear, but could be due to a downturn in the market for spectrum at the time of the auction.

C-Block Auction

The auction of one of the blocks of spectrum allocated for PCS, known as the *C-block*, has presented some complex legal problems for the FCC. In the original C-block auction, also called the entrepreneur's auction, the FCC gave bidding credits to small businesses to help them compete with larger entities in the auction. Winning bidders only had to pay 10% down and the remainder could be paid over ten years at below-market interest rates. Although the initial auction, completed in May 1996, raised $10.2 billion, by mid-1997 many of the license winners (most notably NextWave Telecom, Inc.) defaulted and declared bankruptcy. The licenses were then seized by a court in bankruptcy litigation. In September 1997, the FCC offered a set of options for C-block licensees to restructure their debt (that offer was modified in March 1998). However, the licensees opted to maintain their bankrupt status, preventing the C-block spectrum from being re-auctioned. As a result of a series of decisions in 1999 and 2000 by the U.S. Court of Appeals, the FCC was ultimately able to cancel and re-auction the licenses. The auction for the defaulted licenses was completed January 26, 2001, and yielded $16.86 billion in revenue.

However, on June 22, 2001, the United States Court of Appeals for the District of Columbia found that the FCC did not have the legal right to take back NextWave's licenses for re-auction, and that 216 of the licenses (which

---

[15] FCC CC Docket 92-297 Fourth NPRM and First Report and Order, on Domestic Public Fixed Radio Services, released July 22, 1996, amended by an Order on Reconsideration, released May 16, 1997, and Second Order on Reconsideration, released September 22, 1997 to Establish Rules and Policies for LMDS and FSS.

garnered $15.85 billion in the auction) still belonged to NextWave rather than re-auction winners such as Verizon Wireless. Possible next steps include further litigation or a negotiated settlement between the FCC, NextWave, and auction winners.

To avoid future problems similar to those experienced in the C-block auction, in December 1997 the FCC adopted streamlined auction rules for all services to be auctioned in the future.[16] The rule changes were intended to ensure uniform procedures involving the application, payment, and certain concerns regarding designated entities (i.e., small businesses, women, minorities and rural telephone companies). For example, in many cases the FCC specifies a minimum opening bid prior to an auction, and provides more time prior to the auction for potential bidders to develop business plans, assess market conditions, and evaluate the availability of equipment. The FCC also recommended legislation, which was not enacted (see **Spectrum Auction Procedures** and **Issues for Congress**).

800 MHZ SMR Auction

Another FCC auction that was criticized by some wireless service providers was the auction of SMR licenses in the 800 MHZ range, completed in December 1997. The FCC originally envisioned the 800 MHZ SMR licenses to be similar to those created in the 900 MHZ SMR auctions. The main difference, however, was that many more incumbent SMR licensees existed in the 800 MHZ band than were in the 900 MHZ band. The incumbents were not only concerned about potential interference, but also that they would never again be able to request additional spectrum from the FCC to expand their services. After much contention in a proceeding that lasted three years, the FCC adopted rules for the 800 MHZ licenses, despite the continued dissatisfaction of incumbent SMR licensees. The FCC required incumbents to relocate (against their wishes) to other frequencies after a mandatory negotiation period with new SMR licensees. The new licensees would have to pay for the relocation, but incumbents were forced to compete with the new SMR licensees for the incumbents' existing customers. A total of 524 licenses were sold in the auction, with one large SMR Company, Nextel, winning 90 percent of the new licenses. Many claimed that smaller SMR providers were not able to compete against Nextel in the auction. A

---

[16] FCC 97-413, WT Docket 97-82, ET Docket 94-32, Third Report and Order and Second Further NPRM on Streamlining Auction Rules, released December 31, 1997.

similar set of issues surrounded the FCC's auctions for 220 MHZ licenses, paging services, and location monitoring services.

Incidents of Collusion

In the PCS auction of D, E, and F Block licenses, held in late 1996, some competing bidders were accused of using unusual bid amounts as a means of signaling their market intentions to each other during the auction. By early 1997, the Department of Justice began an investigation into bidding practices employed by participants of the PCS auctions. Based on this investigation, the FCC found specific parties liable for violating FCC auction anti-collusion rules that prohibit bidders from sharing their bidding strategies with competing bidders.[17] The FCC has since modified its bidding procedures so that all bids must be made in specific increments instead of any dollar amount to prevent collusion. It is still possible, however, for bidders to use other forms of collusion to keep prices low.

THE BALANCED BUDGET ACT OF 1997

The Balanced Budget Act of 1997 (47 U.S.C. 153) contained several spectrum management provisions. It amended Section 309(j) of the Communications Act to expand and broaden the FCC's auction authority and to modify other aspects of spectrum management. Whereas previous statutes gave the FCC the authority to conduct auctions, the Balanced Budget Act requires the FCC to use auctions to award mutually exclusive applications for most types of spectrum licenses. Exempted from auctions are licenses or construction permits for:

(A) public safety radio services, including private internal radio services used by state and local governments and non-government entities and including emergency road services provided by not-for-profit organization, that –
    (i) are used to protect the safety of life, health, or property; and
    (ii) are not made commercially available to the public;

---

[17] FCC-98-42, Notice of Apparent Liability for Forfeiture for Facilities in the Broadband PCS in the D, E, and F Blocks. Adopted March 16, 1998.

(B) digital television service given to existing terrestrial broadcast licensees to replace their analog television service licenses; or

(C) noncommercial educational broadcast stations and public broadcast stations.

Examples of services exempted from auctions include utilities, railroads, metropolitan transit systems, pipelines, private ambulances, volunteer fire departments, and not-for-profit emergency road services. This section also extends the FCC's auction authority to September 30, 2007. It directs the FCC to experiment with combinational bidding (i.e., allowing bidders to place single bids on groups of licenses simultaneously), and to establish minimum opening bids and reasonable reserve prices in future auctions unless the FCC determines that it is not in the public interest.

Furthermore, the Act directed the FCC to use auctions for mutually exclusive applications for new radio or television broadcast licenses received after June 30, 199. For applications filed prior to that date, bidding was limited to those who had already filed. Previously, the FCC granted all broadcast licenses through comparative hearing procedures. After a 1993 court case in which FCC criteria for selecting license winners were challenged, however, the FCC had stayed all ongoing comparative hearings pending resolution of the case. Following enactment of the Act, the FCC established auction procedures for all licenses for new commercial radio and television stations, as well as competing applications for new stations filed before July 1, 1997.[18] The FCC has since then conducted several auctions for broadcast licenses and broadcast station construction permits.

The Act directed the FCC to auction 120 MHZ of spectrum, most of which had already been transferred by NTIA from federal to non-federal use. As a result of this provision, the FCC began a proceeding on the allocation and auction of 45 MHZ between 1710-1755 MHZ 9pending), and must allocate by September 2002 another 55 MHZ located below 3 GHz for auction. It also directed NTIA to reallocate another 20 MHZ below 3 GHz (reduced to 12 MHZ be subsequent legislation – see **Allocation of Spectrum for Federal vs. Commercial Use**) for commercial uses. The Act also authorized private parties that win spectrum licenses encumbered by federal entities to reimburse the federal entities for the costs of relocation if the private parties want to expedite the spectrum transfer.

---

[18] FCC 98-194, MM Docket No. 95-31, Implementation of Section 309(j) of the Communications Act – Competitive Bidding for Commercial Broadcast and Instructional Television Fixed Service Licenses, released August 18, 1998.

The Act required the FCC to conduct auctions for 78 MHZ of the analog television spectrum planned to be reclaimed from television broadcasters at the completion of the transition of digital television. That spectrum is to be auctioned in 2002 but not reclaimed from broadcasters until at least 2006. It then directs the FCC to grant extensions to stations in television markets where any one of the following three conditions exist: (1) if one or more of the television stations affiliated with the four national networks are not broadcasting a digital television signal, (2) if digital-to-analog converter technology is not generally available in the market of the licensee, or (3) if at least 15% of the television households in the market served by the station do not subscribe to a digital "multi-channel video programming distributor" (e.g., cable or satellite services) and do not have a digital television set or converter. To maximize the pool of potential bidders in auctions of the returned analog TV spectrum, the FCC may not disqualify bidders due to duopoly or cross-ownership rules if the population of the city in question is greater than 400,000.

Concerning allocation and assignment of new public safety services, the Act directed the FCC to reallocate 24 MHZ between TV channels 60-69 for public safety services and to auction the other 36 MHZ in that band for commercial use. The public safety licenses must be assigned in 1998 and the auction must start by January 1, 2001. During the transition to DTV, the FCC must ensure that new spectrum users and existing television licensees could operate without interfering with each other. The bill directs the FCC to seek to assure that qualifying low power TV stations are reassigned other spectrum where possible.

Furthermore, the Act directed the FCC to allocate spectrum for "flexible use," which means defining new services broadly so that services can change as the telecommunications technology evolves. The FCC was already making such allocations, such as allowing specialized mobile radio services to compete with cellular telephone services, or allowing LMDS to provide two-way communications as well as broadcast services. These allocations must be consistent with international agreements, must be required by public safety allocations and in the public interest, and must not result in harmful interference among users.

Shortly after enactment, the FCC made plans for the required auctions, and later conducted a proceeding to determine which wireless services should be exempted from auctions, to determine the appropriate licensing scheme for new and existing services, and to determine how to implement auctions for services that are auctionable as a result of its revised authority.[19]

STATUS OF SPECTRUM AUCTIONS

To date, the FCC has garnered over $40 billion in total bids from auctioning over 14,300 licenses. According to the FY2002 Bush Administration budget proposal, actual and expected cash receipts total over $31 billion, and spectrum auctions are expected to generate more than $25 billion over the next five years.[20] **Appendix 1** lists types of licenses auctioned to date and an estimate of the expected revenue. As shown, the FCC has conducted auctions for a wide variety of licenses, located at different parts of the radio spectrum, having differing amounts of spectrum, and covering differing geographic ranges. The amounts paid for the licenses depends on these factors as well as many others.

One important measure of the effectiveness of a licensing scheme is the speed with which licenses are granted. Auctions have proven to be far speedier than either comparative hearings or lotteries, cutting the time required to obtain a license from up to four years to under six months. Although auctions have been fraught with a number of problems, few, if anyone, in government, industry, or academia has advocated returning to the use of comparative hearings or lotteries to assign spectrum licenses.

Most observers consider the auctions to be a success, for the federal revenue generated, s well as for the speed with which licenses auctioned have gone to the companies that value them the most and are most likely to put them to use. Moreover, many prefer letting businesses determine whether to invest in a new service rather than relying on the government to decide who receives a spectrum license. The FCC has concluded that auctioning of spectrum licenses has contributed to the rapid deployment of new wireless technologies, increased competition in the marketplace, and encouraged

---

[19] FCC 99-52, TW Docket No. 99-87, RM-9332, RM-9405, Notice of Proposed Rule Making in the Matter of Implementation of Sections 309(j) and 337 of the Communications Act of 1934 as Amended, released March 25, 1999.

[20] FCC rules originally allowed winning bidders to make payments in installments according to license terms. Some auction winners, however, defaulted on their payments, causing a decrease in collections. Installment payments are no longer allowed. See: Congressional Budget Office, The Budget and Economic Outlook: Fiscal Years 2001-2010, Appendix B, CBO Baseline for Spectrum Auction Receipts. January 2000. Also, see: CBO, The Budget and Economic Outlook: Fiscal Years 2002-2011, Box 4-1, January 2001.

participation of small businesses.[21] Many other countries have adopted the use of auctions to assign commercial licenses to use spectrum bands.

RECENTLY COMPLETED AND FUTURE SCHEDULED AUCTIONS[22]

The FCC plans to use auctions to assign licenses for the following other wireless services (summary provided in **Appendix 2**):

PCS C & F Block Re-auction

These are the reclaimed licenses from the defaulted licensees in 1996 C- and F-Block auctions. The FCC made several revisions to the service and auction rules, including reconfiguring the size of the C-Block licenses, modifying auction eligibility restrictions, and retaining the spectrum cap on current spectrum license holders. The licenses are in the frequency range from 1890-1975 MHZ. Two 15 MHZ licenses (paired) and four 10 MHZ licenses (paired) are being auctioned in 196 basic trading area (BTA) markets where licenses were reclaimed, for a total of 422 licenses. Some licenses were open to all bidders, while other licenses were available only to small businesses known as entrepreneurs in "closed" bidding. Some, however, have questioned the FCC's rules that allow small company bidders in the current C-Block auction to obtain funding from large companies, arguing that it fails to help small companies enter the wireless industry.[23] The licenses will be used for broadband PCS, which includes a variety of mobile telecommunications services. The auction began on December 12 and ended on January 26, 2001, raising $16.86 billion in revenue, which is more than any single previous FCC spectrum auction. However, the re-auction was thrown into doubt by the June 22, 2001 decision by the U.S. Court of Appeals of the District of Columbia that the FCC did not have the right to reclaim licenses from firms in bankruptcy (i.e., NextWave), and that

---

[21] FCC 97-353, FCC Report to Congress on Spectrum Auctions, WT Docket No. 97-150, released October 9, 1997.

[22] See [http://www.fcc.gov/wtb/auctions] for additional details on upcoming auctions.

[23] Big Carriers Dominate Cellular Auction, Despite Plan to Help Start-Ups, Wall Street Journal, January 5, 2001, p. B1.

those licenses still belong to those licensees despite defaulting on their payments.

700 MHZ Guard Band

This spectrum, located in the 746-764 MHZ and 776-794 MHZ bands, was part of the reallocation from television channels 60-69 in connection with the transition to digital television (DTV).[24] To protect public safety users in adjacent spectrum from interference by the new service the FCC established two 6 MHZ "guard bands" at the upper and lower ends of the spectrum allocated for auction. Technical and operational restraints on the use of the guard band spectrum are more stringent than restraints on the other 30 MHZ to be auctioned (discussed below). Auctions for licenses in the guard bands are being conducted in two parts. The first part was conducted in September 2000, raising $520 million. The second part (auctioning eight licenses that were not sold in the first guard band auction) was concluded on February 21, 2001 and raised $20.9 million.

700 MHZ Band

The FY2000 Defense Appropriations Act (P.L. 106-79, Title VIII, Sec. 8124) directed the FCC to conduct auctions for licenses in the upper 700 MHZ band (spectrum reallocated from channel 60-69 television services) so that proceeds are deposited in the U.S. Treasury by September 30, 2000. The FCC originally scheduled auctions for this spectrum (including the guard bands) to meet that deadline. Later, however, the FCC requested permission from House and Senate Appropriations Committee Chairmen to delay the auctions to allow bidders to develop "better business plans and bidding strategies and to form strategic alliances." Although no legislation was introduced, support was informally expressed by the Committees of jurisdiction, and the FCC has most recently postponed most of the upper 700 MHZ band auctions (all but the guard bands) until September 2001.[25]

---

[24] Part of this reallocation (the 6 MHz guard bands) was already auctioned in September 2000; the other 24 MHz in those band was reallocated for public safety services.

[25] FCC Memorandum Opinion In the Matter of Cellular Telecommunications Industry Association et al.'s Request for Delay of the Auction of Licenses..., released September 12, 2000.

Spectrum auctioned will be 746-764 MHZ and 776-794 MHZ bands. A total of 12 licenses will be auctioned – one 20 MHZ license (consisting of paired 10 MHZ blocks) and one 10 MHZ license (consisting of paired 5 MHZ blocks) – in six regions known as economic area groupings. These licenses would be considered highly desirable licenses because of their VHF frequency range, except that incumbent television broadcasters are currently using the spectrum, and will continue using it until at least December 31, 2006 (most likely longer). License winners may not cause interference to incumbent broadcasters, making it very difficult to use the spectrum for some time in the more populated parts of the country. The licenses are intended to be used for high-speed Internet access, new fixed wireless operations in under-served areas, and next-generation high-speed mobile services. Accordingly, incumbent channel 60-69 broadcasters have been negotiating a "buyout" with the wireless industry, whereby broadcasters will agree to vacate the spectrum early in exchange for a percentage of the amount wireless companies are expected to pay in auction.[26]

The FCC has scheduled the upper 700 MHZ band auction for September 12, 2001. The Bush Administration's FY2002 budget proposes legislation that would promote clearing channels 60-69 spectrum for new wireless services in a manner that ensure incumbent broadcasters are fairly compensated. The FY2002 budget proposal would delay the auction until September 2004 in order to increase the spectrum's value and garner more revenue for the Treasury. To date, the FCC has not announced any intention to further delay the auction.

Meanwhile, in a separate proceeding, on March 28, 2001, the FCC issued a notice proposing the auction of the lower 700 MHZ band (698-746 MHZ, television channels 52-59) for commercial wireless services. Public comments on this proposal are due to the FCC by May 14, 2001. The Bush Administration has proposed delaying this auction from 2002 to 2006.

FM Broadcast

These auctions will be for construction permits for FM broadcast stations at 352 locations across the country. Licenses are for the use of spectrum in the FM band (88-108 MHZ) and entail the normal interference

---

[26] As a condition for the buyout, some broadcasters are also asking the FCC to require cable companies to carry all of a broadcaster's DTV signals ("multiple carriage"). See: McConnell, Bill, "Paxson Circles the U's," Broadcasting & Cable, February 26, 2001, p. 13.

protection constraints of FM licenses. The auction is scheduled to begin on December 5, 2001.

Paging Services

The FCC has announced plans to auction 14,000 licenses in the lower paging bands (35-36 MHZ, 43-44 HMZ, 152-159 MHZ, and 454-460 MHZ), and 1,514 licenses in the upper paging bands (929-931 MHZ) that remained unsold in the first paging auction of March 2000. Auction of the lower and upper paging bands is scheduled to commence on June 26, 2001.

Auctions Not Yet Scheduled

The FCC has begun proceedings to plan for auctions for the following other spectrum licenses: AM broadcasting licenses for applicants that filed within designated time frames, licenses for fixed wireless services in the 24 GHz band (called digital electronic message service, or DEMS),[27] licenses around 4.9 GHz transferred from federal government to private sector use, additional licenses for narrowband PCS, low power television (LPTV) and translator stations, and the two additional services described below.

218-219 MHZ (formerly IVDS phase 2)

The FCC had planned to offer two IVDS licenses in each of the 428 rural service areas (RSAs), plus 127 MSA licenses on which the previous winning IVDS bidders defaulted. Just prior to the scheduled start date (February 18, 1997), the auction was postponed indefinitely as a result of numerous petitions from industry and Congress to revise the service rules to make these licenses more attractive to bidders. In 1999, the FCC offered financial restructuring to current IVDS licensees, and debt forgiveness to previous licensees who made their first two down payments but did not make their March 16, 1998, payment. The FCC also changed the IVDS service rules to increase the flexibility of licenses to allow the provi9son of Internet services.

---

[27] DEMS licenses will each be 40 MHz (paired).

General Wireless Communications Services (GWCS)

The FCC has allocated five licenses, 5 MHZ each, between 4660-4685 MHZ, to be auctioned in each of 175 Economic Areas, for a total of 875 licenses. There are, however, incumbent licenses in that spectrum in many parts of the country. GWCS licenses, conceived to be similar to WCS, may be used to provide any fixed or mobile communications services except broadcast, radiolocation, and satellite services. These may include voice, video, and data services, private microwave, broadcast auxiliary, and ground-to-air signals. The question of what to do with the incumbents, however, is causing difficulty in designing the auction.

SPECTRUM VALUE

Spectrum value depends on many factors, such as the amount of spectrum, its frequencies (since signal transmission characteristics vary along different parts of the spectrum), the geographic area covered, the services permitted by FCC rules, the availability of equipment that can operate at those frequencies, the demand for services that do not interfere with other bands, the amount of alternative spectrum already available for similar services, the number of incumbents presently occupying the spectrum, and whether incumbents will remain in that spectrum or be relocated to other spectrum. Spectrum value may be greater if adjacent bands can be aggregated to form larger blocks and if the given spectrum is not encumbered by other licensees using the same frequencies. Giving bidders enough time to review the auction rules and services rules, examine technical opportunities and constraints, prepare marketing plans, and arrange financing is also critical to obtaining full value of the auction. It is impossible to determine in advance precisely the revenue that can be obtained from a given spectrum auction.

After an auction closes, spectrum value is often measured by the total dollars raised per "MHZ-pop" (the number of MHZ provided in a license multiplied by the total population covered by the license, similar to a unit price). However, the MHZ-pop value of a given license can vary significantly from one auction to another. In the PCS auctions, for example, the narrowband PCL licenses drew over six times more revenue per MHZ-pop than the broadcast PCS licenses, but drew much less total revenue because of the smaller amount of spectrum auctioned.

The Congressional Budget Office (CBO) annually scores the anticipated receipts from planned spectrum auctions, and includes the revenue estimate

in its annual report, *The Budget and Economic Outlook.*[28] The January 2001
report estimates receipts from the spectrum at nearly $28 billion over the
2001 through 2011 period. The revenue expected from the auctions is used
as offsetting receipts to other federal expenditures. In accordance with the
Budget Enforcement Act of 1990,[29] the auction proceeds, as assessed by
CBO, cannot be used for funding other programs.

TECHNOLOGY INNOVATIONS

Several technological advances could affect the outcome and prospects
for spectrum auctions and how the spectrum is managed. The usable
spectrum for communications purposes is currently considered to be below
300 GHz. Higher frequencies present limitations such as a greater absorption
of signals by the atmosphere, and difficulties in high frequency reception. As
the technology for radio transmission and reception improves, higher
frequencies will likely become available for use. Technology improvements
may, in turn, spur increased consumer demand for spectrum.

Some of the problems with high-frequency signal transmission and
signal interference at all frequencies are being solved by engineering
techniques which could make better use of the spectrum, thus reducing some
of the spectrum demand. These include methods of digital signal
compression, which increases the carrying capacity of currently sued bands,
error detection and correction which maintain the signal integrity even in
high levels of noise, and other digital techniques such as frequency hopping,
in which the transmitted signal avoids frequencies that are already being
used. The use of fiber optic cables (which carry signals over wires rather
than propagating through the air, and therefore do not require frequency
allocations) can provide enormous capacity and alleviate some of the
demand for spectrum. Cables, of course, cannot be used for services that
require wireless transmission. Two other advanced technologies called ultra-
wide band technology and software-defined radio, are discussed under
Recent Developments in Spectrum Management.

In developing service and auction rules for spectrum license auctions,
the FCC tries to maximize spectrum flexibility by allowing licensees to offer

---

[28] CBO does not break down the amounts in terms of individual auctions, but only provides the
aggregate estimated revenue.

[29] For a discussion of the Budget Enforcement Act of 1990 and other budgetary requirements,
see CRS Report RL30363, The Sequestration Process and Across the Board Spending
Cuts for FY2001.

as many services as possible without interfering with existing spectrum use. The FCC also usually considers competitive market forces in allocating and licensing spectrum. This entails using auctions for many new terrestrial spectrum licenses, and defining new services broadly enough to allow services to change as the technology evolves.

In some auctions, spectrum is sold for the same spectrum bands in the same geographic areas as incumbent licensees are located. The new licenses are called overlay licenses because they use frequencies that surround the frequency of an existing license. The auction winner must prevent the operations of its overlay license from interfering with those of an incumbent licensee. The new licensee could either "work around" the spectrum of the incumbent license (by using frequency hopping) or pay for the relocation of the incumbent to some other frequency. Overlay licensees were implemented in the PCS auctions since there were already incumbent licensees (called microwave licensees) using that spectrum. To help clear the PCS spectrum of microwave incumbents, the FCC provided higher frequency spectrum for the incumbents and required PCS licensees to pay for the costs of relocating incumbents to high frequencies. However, placing that requirement on the new licensee typically lowers the value of the license. Most licenses currently auctioned by the FCC are encumbered with existing licensees to an even greater extent that the PCS spectrum. If the cost of relocating the incumbent exceeds the value of the license, it can be difficult to attract bidders in an auction. A further difficulty is that in some auctions, the FCC has not provided new spectrum to relocate incumbents, leaving those negotiations for after the auction.

To help smaller businesses participate in an auction, the FCC sometimes allows license winners to partition licenses into smaller geographic areas than were originally defined by the FCC. This allows a wireless service provider to set up a business in a smaller community without having to serve an entire region. The FCC also allows some licensees to "disaggregate" a portion of their spectrum, i.e., to divide the spectrum into several smaller bands. This enables smaller companies to use a portion of the spectrum for some specialized service.

RECENT DEVELOPMENTS IN
SPECTRUM MANAGEMENT

Intelligent Transportation Systems

The automotive industry and the federal government have been working together for many years developing the electronics, communication systems, and information processing capability to improve the efficiency and safety of surface transportation systems. These planned systems are collectively referred to an intelligent transportation systems (ITS).[30] In October 1999, in accordance with provisions in the Transportation Equity Act for the 21st Century (23 U.S.C. 502),[31] the FCC allocated 75 MHZ in the 5.850-5.925 GHz band for ITS on a co-primary basis with current users of that spectrum. ITS users were given the same level of protection from interference as other primary users (which include federal (primary defense) operations, and commercial fixed satellite services), and a higher level of protection than secondary users (amateur radio). ITS will use this spectrum for dedicated short range communications such as traffic light control, traffic monitoring, travelers' alerts, automatic toll collection, and other purposes. The FCC deferred consideration of licensing and service rules to a later proceeding, in anticipation of further details on ITS requirements from the Department of Transportation.

Technology Advisory Council

In early 1999, with the increasing number of conflicts associated with spectrum management, and the burgeoning wireless communications industry, the FCC established the Technology Advisory Council. The Council, comprised of industry and academic experts, would help the FCC in planning its strategy for regulating the wireless industry, as well as analyze issues of convergence with the Internet and other technical issues. Since its inception, the Council has studied issues of spectrum management,

---

[30] For information on ITS, see CRS Report RL30403, Intelligent Transportation Systems: Overview of the Federally Supported Research and Development Program, January 11, 2000.

[31] The provision directs the Secretary of Transportation, in consultation with the Secretaries of Commerce and Defense and the FCC, to secure the necessary spectrum for the establishment of dedicated short-range vehicle to wayside wireless services.

electromagnetic noise created by interference among commercial and government wireless systems, access to telecommunications by persons with disabilities, network interconnection, and network access.

The Council investigated two technologies in particular, which are thought to have the potential to alleviate some of the demand for spectrum. One, called *ultra wide-band* technology, is an innovation involving the spreading of a radio signal over a wide range of frequencies that are already assigned to other communications services. If the ultra wide-band signal is transmitted at a low enough power level, it would not interfere with existing signals. The other technology, called *software-defined radio*, is the development of a new type of radio equipment that can be quickly reprogrammed to transmit and receive on any frequency within a wide range using any transmission format. The FCC has proposed allowing the limited use of both of these technologies on an unlicensed basis. However, incumbent users of the spectrum that would be used for these technologies are concerned about the FCC proposal. The incumbents (primarily the cellular and PCs industries, and operators and users of the Global Positioning System satellite navigation system) argue that potential interference may cause failures in their systems, some of which involve public safety.

FCC Policy Statement and Regulatory Framework for Auctions

In November 1999, the FCC released a statement outlining guiding principles for its future activities in spectrum management.[32] The statement was intended to provide a framework for industry and government parties to understand future FCC decisions regarding reallocation of the remaining 200 MHZ of spectrum to be transferred from government to non-government radio services as per statutory requirements. The new strategy contained the following principles:

- allow greater flexibility in allocations, including "harmonization" of FCC service rules to provide regulatory neutrality for similar wireless services;

---

[32] FCC 99-354, Policy Statement. In the Matter of Principles for Reallocation of Spectrum to Encourage the Development of Telecommunications Technologies for the New Millennium, released November 22, 1999.

- promote new spectrum-efficient technologies, such as ultra-wideband and spread spectrum operations;

- ensure that important communications needs, such as public safety, are met;

- improve the efficiency of assigning licenses through streamlining and innovative techniques;

- encourage the development of secondary markets for spectrum (i.e., reselling of licenses to third parties) to ensure full utilization; and

- seek ways to make more spectrum available, through, for example, assigning user fees or by reclaiming existing spectrum.

One innovation that was introduced to improve the efficiency of assigning licenses is the "band manager" concept. Under this approach, licenses for blocks of spectrum would be auctioned to band managers, who would then subdivide and lease portions of their spectrum in response to market demand. Prices charged to users would be set by competition among the band managers for potential spectrum users.

The Policy Statement then inventoried spectrum that was available for allocation and outlined proposals for allocating and assigning that spectrum. At the same time, the FCC created a Spectrum Policy Executive Committee to address policy issues affecting spectrum management, to implement initiatives consistent with the Policy Statement, and to coordinate related actions among the FCC's internal bureaus.

In November 2000, the FCC established a regulatory framework for future auctions, which provided further details on its spectrum management plans, and answered many questions raised in the proceeding it initiated in March 1999.[33] In the framework, the FCC may conduct auctions for licenses for private radio services to resolve mutually exclusive applications if the FCC "determines that it is in the public interest to do so." The FCC will continue to decide on a service-by-service basis the licensing scheme for new services. One option discussed is to use a "band manager" concept (described above) which has been used in the 700 MHZ Guard Band auction in September 2000. In addition, the ruling defines the scope of the statutory exemption from auction for public safety radio services to include not only police, fire, and emergency medical services, but also non-commercial

---

[33] FCC 00-403, WT Docket No. 99-87, Report and Order and Further Notice of Proposed Rule Making (NPRM), released November 20, 2000.

services used by entities such as utilities, railroads, and transit systems. The ruling also addressed a number of pending petitions to amend its licensing and eligibility rules for existing private wireless services (see **Issues for Congress, Private Land Mobile Radio Services**).

Recent Spectrum Allocations

In October 2000, the FCC allocated 50 MHZ (3650-3700 MHZ, which had previously been transferred from government use) for commercial wireless services.[34] The FCC intends to permit and encourage the use of this spectrum for new broadband, high-speed wireless voice and data services, particularly in rural areas. The spectrum will continue to be encumbered by existing licensees in the fixed satellite services, which will share the band with new licensees. The FCC proposed to assign the new licenses through auctions.

In November 2000, the FCC proposed to reallocated 27 MHZ of spectrum, previously transferred from federal government use, to various non-government services.[35] The spectrum, located in a number of separate bands, could be allocated for private land mobile services, satellite feeder links, utility telemetry to support meter reading, and personal location services. Some of the licenses could be assigned to band managers through auctions.

In another November 2000 action, the FCC proposed to adopt a new policy to promote the development of "secondary markets" in radio spectrum.[36] The FCC Policy Statement articulates its goal of promoting a system of secondary markets (i.e., the sale and lease of the right to use spectrum by licensees) to better utilize spectrum that is already licensed. The FCC proposed to allow wireless radio services licensees, with exclusive rights to their assigned spectrum, to lease their spectrum rights to third parties without having to secure prior FCC approval.

In January 20001, the FCC released a proposal to consider possible uses of several frequency bands below 3 GHz for new advanced wireless systems,

---

[34] FCC 00-363, Docket No. 98-237, First Report and Order and Second NPRM, announced October 12, 2000.

[35] FCC 00-395, ET Docket no. 00-221, NPRM In the Matter of Reallocation of...Government Transfer Bands, released November 20, 2000.

[36] FCC 00-041, Policy Statement, released December 1, 2000, and FCC 00-402, WT Docket 00-230, NPRM Promoting Efficient Use of Spectrum Through Elimination of Barriers to the Development of Secondary Markets, released November 27, 2000.

including third generation (3G) systems (see discussion of 3G spectrum).[37] Portions of the 1710-1850 MHZ and 2110-2165 MHZ bands (previously transferred from federal government to non-government use) could be allocated for 3G services, and various approaches for using the 2500-2690 MHZ band will be considered. In the same action, the FCC adopted an Order denying a petition by the Satellite Industry Association for parts of the 2500-2690 band to be reallocated to mobile-satellite services.

---

[37] FCC 00-455, ET Docket No. 00-258, NPRM and Order In the Matter of Amendment of Part 2 of the Commission's Rules to Allocate Spectrum Below 3 GHz for Mobile and Fixed Services to Support..., released January 5, 2001.

Chapter 4

ISSUES FOR CONGRESSIONAL CONSIDERATION

ALLOCATION OF SPECTRUM FOR FEDERAL VS. COMMERCIAL USE

Tension, which has always existed between federal agencies and the private sector over spectrum allocations, has been increasing in recent years. The Omnibus Budget Reconciliation Act of 1993 (47 U.S.C. 927) directed federal agencies to vacate 200 MHZ of spectrum below 3 GHz for reassignment to commercial uses. NITA, manager of all spectrum used by the federal government, identified 235 MHZ to transfer to the FCC,[38] but claimed that releasing any additional spectrum could result in costs greater than the potential revenue from an auction, and could compromise national security, public safety, law enforcement, and air traffic control operations.[39] Nevertheless, a provision in the FY1997 Omnibus Appropriations Act (P.L. 104-208) directed the FCC to auction 30 MHZ of spectrum previously allocated for shared commercial and government radio services (included as part of the 235 MHZ identified by NTIA). The Department of Defense (DOD), the largest federal user of spectrum, was the most vocal of the federal agencies protesting the transfer of spectrum away from federal use.

---

[38] The amount of spectrum identified by NITA was 35 MHz greater than required. This was possibly in anticipation of future demand for commercial spectrum.

[39] Testimony of Hon. Larry Irving to House Commerce Committee, Subcommittee on Telecommunications, Finance, March 21, 1996, and to Senate Commerce Committee, June 20, 1996.

DOD argued that it needs all of the spectrum it is currently assigned to maintain high quality communications to support national security.

Amid the protests by NTIA, DOD, and other federal agencies, the commercial sector increased its pressure on Congress to release additional federal spectrum for commercial use. The Balanced Budget Act of 1997 (P.L. 105-33) directed NTIA to reassign an additional 20 MHZ below 3 GHz to the FCC for auction. In 1998, NTIA identified 20 MHZ (located in five spectrum bands) and a schedule for reallocation from federal agency use within ten years. The report concluded, however, that such a spectrum release could adversely affect critical agency missions and the ability to provide services to the public.[40] The report estimated the cost to federal agencies to be over $1 billion to modify existing equipment and facilities to use alternative frequencies, assuming that suitable spectrum will be available, based on the assumption that extensive system modifications would not be required to avoid interfering with new commercial users. The report stated that the loss of the identified spectrum could restrict spectrum use during defense training exercises, ultimately affecting operational readiness and national security.

In two follow-on reports, NTIA warned that reallocating this spectrum would have a "profoundly negative impact on the planned U.S. space program" and other federal systems, and identified alternative bands to be considered for auction and use for new wireless services.[41] The alternate spectrum included non-federal bands, bands that NTIA had previously released to the FCC, and a shared band between federal and non-federal uses. The telecommunications industry, criticized the NTIA's selection, claiming that the alternative bands were undesirable for commercial use and were encumbered by federal users. In October 2000, the FCC reallocated the bands recommended by NTIA for commercial services, and proposed to assign the licenses through the use of auctions.[42]

While the FCC has reallocated many of the bands identified by NTIA for transfer to non-federal uses, several bands must still be reallocated and licensed. **Table 1** provides a list of the spectrum bands identified by NTIA for transfer to commercial uses, but which either have not yet been reallocated or have not yet been licensed to commercial services. The table

---

[40] NTIA Publication 98-36, Spectrum Reallocation Report: Response to Title III of the Balanced Budget Act of 1997, released February 1998.

[41] NTIA Publication 98-37, Reallocation Impact Study of the 1990-2110 MHz Band, and NTIA Publication 98-39, Identification of Alternate Bands: Response to Title III of the Balanced Budget Act of 1997, both released November 1998.

[42] FCC 00-363, Docket No. 98-237, First Report and Order and Second NPRM, released October 24, 2000.

indicates the current allocations for each of the bands, the status of the transfer of the bands to non-federal uses, and statutory auction deadlines, where applicable. Congress will likely continue to be pressured by the wireless industry to expedite the FCC's reallocation and licensing of these bands for commercial use. Another issue in which Congress could become involved is how to compensate federal agencies for relocating their wireless services to new frequencies. NTIA is planning to begin a proceeding on that issue.

In addition, two legislative measures enacted in recent years will affect the transfer of spectrum from federal to non-federal uses. The FY1999 Defense Authorization Act (47 U.S.C. 923, Title X, Sec. 1064) requires any entity that purchases a license for spectrum previously reserved for used by a federal agency to reimburse that agency for the costs incurred by the agency in relocating its communications to other frequencies. Previously reallocated spectrum and reallocations n the 1710-1755 MHZ band are exempt from this reimbursement. This provision will have the effect of lowering the value of the spectrum at auction, and could cause delays in the licensing process if there are disputes between federal users and license winners over the costs of relocation.

Table 1. Spectrum to be Reallocated from Federal to Non-Federal Uses

Spectrum Band (MHZ)	Primary Current Allocations	Status of Reallocation	Statutory Auction Deadline
216-220	Maritime mobile, inter-active video and data service (218-219 MHZ)	FCC Rulemaking in progress (FCC 00-395)	9/30/2002 (BBA97)
1390-1395	Fixed, radiolocation	FCC Rulemaking in progress (FCC 00-395)	None
1395-1400	Wireless Medical Telemetry Service	Reallocated by FCC to Wireless Medical Telemetry Service*	None
1427-1429	Space operation, fixed mobile	FCC Rulemaking in progress (FCC 00-395)	None
1429-1432	Wireless Medical Telemetry Service	Reallocated by FCC to Wireless Medical Telemetry Service* FCC Rulemaking in progress (FCC 00-395)	None
1432-1435	Fixed, mobile	FCC Rulemaking in progress (FCC 00-395)	9/30/2002 BBA97
1670-1675	Meteorological satellites, meteorological aids	FCC Rulemaking in progress (FCC 00-395)	None
1710-1755	Fixed, mobile	FCC Rulemaking in progress (FCC 00-455) Released January 5, 2001	9/30/2002 (BBA97)
2300-2305	None	FCC set aside as reserve spectrum	None
2385-2390	Mobile, radiolocation	FCC Rulemaking in progress (FCC 00-395)	9/30/2002 (BBA97)
2400-2402 2417-2435	None	FCC set aside as reserve spectrum	None
3650-3700	Fixed satellite, radio-location, aeronautical radionavigation	FCC Rulemaking in progress (FCC 00-363)	None
4940-4990	Fixed, mobile	Substituted by the President for the originally identified 4635-4685 MHZ band. FCC Rulemaking in progress (FCC 00-363)	None

Source: CRS, based on FCC and NTIA data
OBRA93=Ommibus Budget Reconciliation Act of 1993, BBA97=Balanced Budget Act of 1997
* These bands will be shared on a co-primary basis with existing government operations (FCC 00-211 June 2000).

The FY2000 Defense Authorization Act (P.L. 106-65, Title X, Sec. 1062, *Assessment of Electromagnetic Spectrum Reallocation*), directed the Department of Commerce (DOC) and the FCC to conduct a review and assessment of national spectrum planning; the reallocation of federal spectrum to non-federal use in accordance with existing statutes; and the implications for such reallocations to the affected federal agencies. Particular attention was to be given to the effect of the reallocations on critical military and intelligence capabilities, civilian space programs, and other federal systems used to protect public safety, as well as future spectrum requirements of federal agencies. In response to this requirement, as part of the review and assessment of spectrum planning, in November 2000 the FCC requested comments from industry on procedures for reimbursement of relocation costs to federal spectrum users.[43] In January 2001, NTIA proposed rules governing reimbursement to federal entities by the private sector related to reallocation of spectrum.[44] Final rulings by both the FCC and NTIA are expected later this year.

The FY2000 Defense Authorization Act further states the DOD is not required to transfer any spectrum bands to the FCC unless NTIA, in consultation with the FCC, makes available an alternative band or bands to DOD. Further, DOD, DOC, and the Joint Chiefs of Staff must certify that the alternative band (or bands) provides comparable technical characteristics to maintain essential military capability. The Act further requires that 8 MHZ (located in the following three bands: 139-140.5 MHZ, 141.5-143 MHZ, and 1385-1390 MHZ) that were identified for auction in the Balanced Budget Act of 1997, be reassigned to the federal government for primary use by DOD. The conference report (H.Rept. 106-301) urges DOD to "share such frequencies with state and local public safety radio services, to the extent that sharing will not result in harmful interference between DOD systems and the public safety systems proposed for operation on those frequencies." Those 8 MHZ were subsequently reclaimed by the President for exclusive federal use.

The provision in the FY2000 Defense Authorization Act reclaiming 8 MHZ of spectrum for DOD use had an impact on two reallocation proceedings. First, the FCC was developing a plan for the use of part of the 8 MHZ (the 138-144 MHZ band) for interoperable communications among federal, state, and local public safety wireless systems. Another band of

---

[43] FCC 00-395, ET Docket No. 00-221, NPRM In the Matter of Reallocation of Government Transfer Bands, released November 20, 2000.

[44] NTIA Docket No. 001206341-01, RIN 0660-AA14, Notice of Proposed Rule Making posted in Federal Register January 18, 2001.

spectrum will have to be found for that purpose.[45] Second, in February 1998, in fulfillment of the Balanced Budget Act of 1997, NTIA identified another part of the 8 MHZ (139-143 MHZ) for the FCC to reallocate and assign by auctions. Since the FY2000 Defense Authorization Act supercedes previous laws, that spectrum will not be transferred, causing a shortfall in expected revenue from spectrum auctions in FY2002. The Congressional Budget Office (CBO), however, estimated the budget impact of foregone spectrum receipts due to this provision to be $500 million or less. This low estimate was due, in part, to the requirement that spectrum auction winners reimburse the federal agencies for the costs associated with relocating to new frequencies. Congress might decide to monitor the implications of these laws and related actions on future reallocations of spectrum for federal or commercial uses. The General Accounting Office is currently investigating this issue.

SPECTRUM AUCTION AND
LICENSE PAYMENT PROCEDURES

As a result of a number of problems associated with the spectrum auctions (in particular the C-Block auction), in October 1997 the FCC recommended to Congress the following legislative actions:[46]

- to clarify that FCC licensees who default on their installment payments may not use bankruptcy litigation to refuse to relinquish their spectrum licenses for re-auction;

- to grant the FCC explicit statutory authority to manage its installment payment portfolio flexibly;

- to exempt auction contracts from certain provisions of the Federal Acquisitions Regulations (FAR); and

- to modify the statue of limitations for forfeiture proceedings against non-broadcast licensees from one to three years.

---

[45] FCC WT Docket 98-86, First Report and Order and Third NPRM, Development of Operational, Technical, and Spectrum Requirements for Meeting Federal, State, and Local Public Safety Agency Communication Requirements Through the Year 2010, released September 29, 1998.

[46] FCC 97-353, FCC Report to Congress on Spectrum Auctions. WT Docket No. 97-150, released October 9, 1997.

In addition, the FCC streamlined its rules to simplify the auction process (e.g., the applications and payment procedures for bidders), and adopted uniform affiliation rules and ownership disclosure rules to avoid legal problems associated with a 1995 Supreme Court decision limiting special treatment for women-owned, and minority-owned companies.[47] The revised rules also provide for higher bidding credits for small businesses (15, 25, and 35 percent, based on the size of the business).

Legislation was introduced in 1998, and again in 1999, addressing these concerns. The 1999 provision was in a section of the Senate version of the FY2000 Appropriations bill for the Departments of Commerce, Justice, State, and Related Agencies (S. 1217, Sec. 618, introduced June 14, 1999) that would have met some of the FCC's requests. The provision authorized the FCC to recover and re-auction licenses if a license fails to meet its installment payment obligations; it allowed the FCC to avoid the jurisdictional and administrative burden associated with reclaiming a license under state laws; and the provision was retroactive to include pending cases. Several wireless service providers opposed the provision, and it was removed in conference.

The provision on auction procedures was not included in FY2001 Appropriations or any other legislation introduced in 2000. Some of the motivation for the legislation has subsided since the FCC now requires payment in full by license winners. Moreover, a series of court decisions over the past several years involving the defaulted C-block licensees has enabled the FCC to proceed with the re-auction of these licenses. However, in the unlikely event that a higher court overturns the latest ruling, winners of some of these licenses will have to give them back.[48]

The FCC continues to express a need for a provision to establish its regulatory authority over spectrum licenses in all states and jurisdictions in the country.[49] Some companies in the wireless industry continue to oppose such a provision, while other companies advocate it. An issue before the 107th Congress is whether to review the FCC's recommendations in light of all of the recent changes made in spectrum management policy to determine whether a legislative remedy is warranted.

---

[47] See Footnote 13.

[48] E-Business Auction: Airwaves Auction Pulls in 16.86 Billion, Wall Street Journal, January 29, 2001, p. B4.

[49] The court decision giving the FCC the authority to re-auction the licenses to defaulted C-Block licensees only applied to the legal jurisdiction that includes New York, where the defaulted licensees (most notably NextWave Inc.) were headquartered.

DIGITAL TELEVISION SPECTRUM

Digital television (DTV) is a new television broadcasting service developed by the television and computer industries. Without the authority to conduct auctions for DTV licenses, the FCC granted, free of charge, licenses to broadcasters for DTV transmissions, as directed by the Telecommunications Act of 1996 (47 U.S.C. 336). In April 1997 the FCC granted all full power television broadcasters (over 1600 stations), as part of their licenses, 6 MHZ to transmit DTV programming over the air in addition to their 6 MHZ of analog television spectrum.[50] The FCC and the television industry are planning a transition from the current analog television broadcasting system to DTV over the next several years. The FCC's plan is that at the end of the transition to DTV, the broadcasters will return the 6 MHZ currently used for analog television broadcasting. Some in industry, the FCC, and Congress are concerned that the transition is not proceeding rapidly enough, and that DTV might fail to become competitive with other multi-channel video services.

During the transition to digital television, broadcasters are transmitting both the existing analog television and DTV signals, so that consumers can continue to receive television programming on their existing receivers until they purchase new digital television sets or converters. Television stations in the nation's top ten markets started to provide a digital signal on November 1, 1998. All remaining commercial DTV stations are to be constructed by May 1, 2002, and non-commercial DTV stations by May 1, 2003. The FCC had originally set 2006 as the target date for broadcasters to cease transmitting the analog signal and return the 6 MHZ of analog television spectrum to the FCC to use for other purposes. The Balanced Budget Act of 1997 (P.L. 105-33), however, prohibited the FCC from reclaiming the analog spectrum from a broadcaster if at least 15 percent of the television households in that broadcaster's market have not purchased DTV receivers or converters. Most analysts believe it is unlikely that most markets will have met that criterion by the 2006 target date. Broadcasters in some markets may not have to surrender their spectrum for many more years. Some question whether the FCC will ever be able to reclaim the spectrum. In aggregate, the auctionable analog TV spectrum represents a large amount (about 84 MHZ) of highly desirable spectrum.[51] However, the uncertainty

---

[50] FCC Fifth Report and Order on Advanced Television Systems and Their Impact on Existing Television Service, released April 21, 1997.

[51] See Completing the Transition to Digital Television, Congressional Budget Office, September 1999. page 14.

over the availability of this spectrum will present difficulties for the FCC in attempting to attract bidders to an auction for the spectrum.

DTV services currently consist only of high definition television (HDTV) during prime-time programming hours, but broadcasters are also planning to use their DTV spectrum during other parts of the day to simultaneously transmit multiple television programs in digital format instead of a single HDTV broadcast. The DTV spectrum can also be used to provide interactive television, Internet access, subscriptions to multimedia news services, and other information services. Each new service provides new options for consumers and new revenue to broadcasters. Currently, over 500 DTV stations have been granted construction permits, and over 180 are on the air. Despite criticism from observers over the slowness of the transition to DTV, the FCC has stated that most of the delays in construction are a result of matters beyond the broadcasters' control.[52]

Several issues could cause delays in the transition to DTV and in the return of the analog television spectrum. One issue is whether Congress should amend the Balanced Budget Act of 1995 to require broadcasters to return the analog spectrum under stricter deadlines. Another concern is over the adequacy of the FCC's rules requiring fees from broadcasters who use DTV licenses for subscription services.[53] Another involves what public interest requirements should be placed on DTV licensees. Controversy continues between state and local authorities and broadcasters (and other wireless service providers) over the placement and construction of DTV transmission towers. The FCC has not yet determined all of the conditions under which cable TV operators should be required to provide the DTV programming of broadcast TV stations (i.e., the "must carry" debate). Disagreements also continue among cable service providers, television manufacturers, and broadcasters over a standard for connecting DTV sets with digital cable systems. Finally, the transition to DTV depends on the willingness of consumers to purchase DTV receivers or converters. For each of these issues, Congress will likely have to decide to what extent, if any, it should intervene in the development of markets and regulations.[54]

Meanwhile, the Bush Administration's FY2002 budget has proposed "squatting fees" as an incentive for broadcasters to surrender their analog

---

[52] Statement of Dale Hatfield, Chief, FCC Office of Engineering and Technology, before the House Commerce Committee, Subcommittee on Telecommunications, Trade and Consumer Protection, July 25, 2000.

[53] Consumer groups would like the fees to be higher for commercial broadcasters, while the broadcasters think the fees should be lower.

[54] For further analysis of these issues, see CRS Report 97-925, Digital Television: Recent Developments and Congressional Issues, updated February 2, 1999.

spectrum. Under this proposal, the broadcasters would pay the U.S. Treasury $200 million per year in analog spectrum fees from 2002 till 2006. Between 2006 and 2010, fees would be reduced until, it is assumed, all the analog spectrum will have been reclaimed. While similar "squatting fees" were proposed by the previous two Administrations, Congress has not implemented nor endorsed this approach.

LOW POWER FM RADIO SERVICE

In response to numerous inquiries from religious and other local groups, in January 2000, the FCC adopted rules for a low power FM radio (LPFM) broadcasting service to be licensed to local communities. The service consists of two classes of LPFM radio stations with maximum power levels of 10 watts and 100 watts. The rules contain interference protection criteria to help ensure that the LPFM service protects and preserves the technical integrity of existing radio service. Since the inception of LPFM, however, incumbent full-power FM broadcasters and radio manufacturers have protested the FCC ruling. The main argument against LPFM is based on concerns over interference with existing FM radio broadcasts, and the potential that LPFM might impede the future transition to digital audio broadcasting.

Many Members sought to severely scale back or nullify the FCC's decision to issue LPFM licenses, while other Members supported the FCC ruling. On April 14, 2000, the House passed the Radio Broadcasting Preservation Act of 2000 (H.R. 3439, as amended), to prevent LPFM licenses from being located within three FM channels away from incumbent full-power broadcasters, while directing an independent study of whether LPFM is causing harmful interference to full-power broadcasters. Three bills were introduced in the Senate: S. 2068, to prohibit the FCC from authorizing LPFM licenses; S. 2518 (later introduced in modified form as S.2989), to permit a limited number of LPFM licenses; and S. 3020, similar to the House-passed bill. Language similar to the House-passed bill was inserted into the FY2001 Appropriations Bill for the Departments of Commerce, Justice, and State, and related agencies (H.R. 4690), which was not signed by President Clinton, citing the LPFM provision as one of the issues that must be resolved.[55] The provision was again included in the District of

---

[55] For further discussion and analysis of the LPFM issues, see CRS Report RL30462, Low Power FM Radio Service: Regulatory and Congressional Issues, updated regularly.

Columbia Appropriations bill (H.R. 4942), (Conf. Rept. 106-1005 contained the Commerce, Justice, and State Appropriations bill, H.R. 5548) that passed Congress on December 15, 2000. It is estimated that the legislation will have the effect of eliminating 75 percent of the potential LPFM licenses that would otherwise be granted.

The 107th Congress might decide to monitor the results of the required study. If it is found that LPFM does not cause harmful interference to full power broadcasters, Congress could reinstate the FCC's original LPFM program.

THIRD GENERATION (3G) MOBILE WIRELESS SERVICES

Rapid growth in the number of subscribers of mobile wireless telecommunications services in the United States and abroad is fueling interest and developments in the next generation of wireless technology services, known as 3G. In addition to the existing wireless communications capabilities, 3G services might include high-speed mobile Internet access and the ability to use the same handset anywhere in the world. Issues related to the implementation of 3G centers mainly on the allocation of spectrum and adoption of technology standards by each of the countries developing this new service. While some steps have been taken to coordinate these activities, much work remains before 3G services will be available to the American public.

The International Telecommunication Union (ITU), a United Nations (UN) agency, is sponsoring the adoption of 3G standards and the allocation of spectrum to integrate various satellite and terrestrial mobile systems into a globally interoperable service.[56] The ITU conducts World Radiocommunication Conferences (WRCs) every two to three years to reach consensus among member states on spectrum allocations. One of the key issues at WRC-2000 was the identification of global spectrum bands that could meet the additional spectrum requirement for 3G services. Once the spectrum bands for 3G were identified internationally, each country had to decide what frequencies within those bands to use for the initial

---

[56] In mobile terrestrial systems, the individual handsets send and receive the telecommunications signals to and from nearby ground-based stations that connect to the public switched telephone network. In mobile satellite systems, the handsets send and receive signals to

implementation of 3G services, as well as the long-term expansion of those services.

One of the identified bands, 1755-1850 MHZ, is currently allocated in the United States for exclusive government use. While the EU would like that spectrum to be allocated for 3G services in the United States, some federal agencies, particularly the Department of Defense (DOD), are concerned that any 3G services that are licensed in that band could interfere with existing communications. Similarly, another band identified for 3G at WRC-2000, 2500-2690 MHZ, also has incumbent licenses in the United States. Incumbents include multi-channel multipoint distribution systems (MMDS, a commercial "fixed wireless" service originally used for television broadcasts, and now being developed for mobile wireless broadband applications), and Instructional Television Fixed Services (ITFS, similar to MMDS but licensed for educational programming).

The need to expedite spectrum 3G allocations was highlighted on October 13, 2000, when a Presidential Memorandum issued by President Clinton directed all federal agencies to work with the FCC and the private sector to identify the spectrum needed for 3G services. The FCC, in conjunction with NTIA, was directed to identify suitable 3G spectrum by July 2001, and to auction licenses to competing applicants by September 30, 2002. Both the FCC and NTIA were tasked with conducting studies into the potential for allocating 3G spectrum. NTIA's final report (*The Potential for Accommodating Third Generation Mobile Systems in the 1710-1850 Band*, March 2001) found that full-band sharing is not feasible because of signal interference with DOD systems, and that relocating DOD spectrum to another band would not be possible, if at all, until beyond 2010.[57] The NTIA has suggested that some limited band sharing options might be feasible if 3G operations can be restricted in space or time, and if 3G operators reimburse certain federal operators to relocated to new frequencies prior to commencing operations near those federal operations. On January 18, 2001, NTIA released a Notice of Proposed Rule Making (NPRM) on the reimbursement procedures associated with the use of that band, and will issue a final rule later this year (for further details see NTIA's 3G website [http://www.ntia.doc.gov/ntiahome/threeg/index.html].

---

and from orbiting satellites that relay the signals to a single ground station that serves a large geographic area, and then connects to the public switched telephone network.

[57] On October 30, 2000, DOD released a report on the electromagnetic compatibility interactions between major DOD radiocommunications systems operating in the 1755-1850 MHz band and potential 3G systems. The report stated that the band is heavily used by government users.

The FCC's final report (Spectrum Study of the 2500-2690 MHZ Band, March 30, 2001) found that band sharing between 3G systems and incumbent licensees (primarily schools with instructional television licenses (ITFS) and fixed wireless broadband providers) would be problematic, and that there is no readily identifiable alternate frequency and that could accommodate a substantial relocation of the incumbent operations in the 2500-2690 MHZ band. In December 2000, the FCC released an NPRM proposing to allocate portions of the 1710-1850 MHZ and 2110-2165 MHZ bands (previously transferred from federal government to non-government use) for 3G services, and seeking comment on various approaches for using the 2500-2690 MHZ band. The FCC also adopted an Order denying a petition by the Satellite Industry Association for parts of the 2500-2690 band to be reallocated to mobile-satellite services (for further details see the FCC's 3G website [http://www.fcc.gov/3G/]).

With the July 2001 deadline approaching for 3G spectrum allocation decisions, on June 26, 2001, FCC Chairman Michael Powell sent a letter to Secretary of Commerce Donald Evans recommending that the deadline be extended "to allow the Commission and Executive Branch to complete a careful and complete evaluation of the various possible options for making additional spectrum available for advanced wireless services." Chairman Powell recommended extension of statutory deadlines as well, stating that "[t]ogether with the Executive Branch and our congressional authorizing and appropriations committees, I expect that we could come up with a revised allocation plan and auction timetable."[58]

PUBLIC SAFETY SPECTRUM NEEDS

Efforts by government, industry, and public safety groups to replace outdated wireless communications systems for public safety agencies is also proving to be a challenge. The main obstacle in these efforts are (1) the high costs of new equipment, (2) the scarcity of unused spectrum, and (3) the need to coordinate among many organizations to enable public safety personnel to communicate with counterparts in other jurisdictions and government levels (known as "interoperability"). While progress has been made by public safety agencies, some argue that given the advances in technology, they should be closer to their goal. Some also compare the

---

[58] Letter from Michael Powell to Donald Evans, available at: [http://www.fcc.gov/Speeches/Powell/Statements/2001/stmkp127.pdf].

developments achieved by commercial wireless services to the status of public safety systems in terms of interoperability and ease of use, and argue that public safety systems should be further along.

Several legislative options could possibly expedite the development of a more unified public safety communications system. One possibility is to direct an increase in coordination among organizations working on public safety wireless communications issues, although introducing new bureaucratic requirements could cause delays. To provide spectrum for these systems, H.R. 4146 (Nick Smith, introduced March 30, 2000), contained a provision directing the FCC to allocate the spectrum between 139-140.5 MHZ, and between 141.5-143 MHZ, inclusive, to interoperability use by public safety services. There was no action on this bill. However, the 107th Congress could pursue legislation to make spectrum that has been reallocated for public safety available at a specified date, although incumbent television broadcasters using that spectrum would likely oppose such a measure. Additional spectrum for public safety may be found through sharing with DOD, as directed by the FY2001 Defense Authorization Act (P.L. 106-398). That law directs DOD, in consultation with the Departments of Justice and Commerce, to identify a portion of the 138-144 MHZ band to share in various geographic regions with public safety radio services.

Regarding funding, for FY2001, $177 million has been provided for the Department of Justice to purchase new communications systems that will interoperate with state and local public safety systems. Other legislation to provide indirect funding for state and local wireless communications systems was introduced, but not enacted. Direct federal funding to state and local public safety agencies for new wireless communications systems would probably require significantly greater appropriations.[59]

In addition to efforts to enable public safety officials to communicate more effectively with each other, there is also an effort in government and industry to enable mobile telephone users to communicate with the public safety agencies in emergencies. This issue was addressed by the Wireless Communications and Public Safety Act of 1999 (P.L. 106-81, enacted October 26, 1999). While this law does not require any new spectrum allocations, it does require wireless service providers to deploy technologies that enable public safety officials to monitor the location of customers. Many are concerned that this requirement could violate the privacy rights of individuals, especially if third parties gain access to the location information

---

[59] For further discussion and analysis of this issue, see CRS Report RL30746, Wireless Communications Systems and Public Safety, November 27, 2000.

of individuals. Some call for legislative or regulatory remedies to prohibit wireless service providers from releasing the location information.

Another public safety spectrum issue is interference in the 800 MHZ band between public safety radio communications and commercial mobile radio services. There has been an increasing number of reports of police and firefighter radios failing to function when used near commercial cell towers and base stations. While all providers are operating within the parameters of their FCC licenses, the FCC convened a working group of interested parties to discuss and study the issue. In February 2001, the working group and the FCC released a "Best Practices Guide" for avoiding interference between public safety wireless communications systems and commercial wireless communications systems for 800 MHZ. The FCC continues to collect information on this problem, and has stated that additional facts and analyses are needed to conclusively establish the causes of this interference and to identify potential remedies.

SPECTRUM CAP

To maximize competition in the wireless industry, in 1994 the FCC established rules to limit the amount of commercial mobile radio services (CMRS) spectrum any one entity can hold in any one market. Accordingly, no licensee in the cellular, PCS, or SMR services is allowed to have an attributable interest in more than 45 MHZ of CMRS spectrum within an urban geographic area. There is currently 180 MHZ licensed to these three services that is subject to the above "spectrum cap." In 1999, the FCC reviewed its spectrum cap policy, and decided that the spectrum cap continues to prevent any one entity from obtaining the majority of spectrum in a market, or warehousing spectrum for the purposes of shutting potential competitors out of a market.[60] The FCC ruled to maintain the cap with some modifications, such as allowing passive investors greater ability to fund small carriers. The ruling also raised the cap to 55 MHZ within rural geographic areas, and allowed for some waivers to be made to the spectrum cap (e.g., in the planned 700 MHZ band auction).

Many wireless service providers criticize the spectrum cap for preventing growth and innovation of their networks. They point to the higher

---

[60] FCC 99-, Docket No. 98-205, Report and Order In the Matter of 1998 Biennial Regulatory Review Spectrum Aggregation Limits for Wireless Telecommunications Carriers, released September 22, 1999.

spectrum caps in other nations (e.g., Japan and Britain), arguing that this has caused the U.S. wireless penetration rates to be lower than those of other industrialized countries. They further argue that even if the spectrum cap was justifiable in the past, it is no longer necessary given the strong competition for wireless services today. Some wireless service providers are reaching capacity with their current spectrum, leading to network congestion, busy signals, and delays for consumers. They argue that the best way to relieve congestion and continue their growth into new wireless services (such as 3G) is to repeal the spectrum cap.

In the 106th Congress, legislation was introduced November 16, 1999 (S. 1923), to prohibit the FCC from applying commercial mobile radio service spectrum aggregation limits (caps) to spectrum assigned by initial auction. This and a similar bill (H.R. 4758, introduced June 26, 2000) would have enabled incumbent mobile service operators to bid for 3G spectrum licenses. At a July 19, 2000, hearing of the House Commerce Committee, Subcommittee on Telecommunications, Trade and Consumer Protection, several witnesses from the wireless telecommunications industry recommended repealing the spectrum cap. The FCC, along with some wireless service providers, argued that the spectrum cap has produced the high level of competition that exists today and needs to be maintained. No action was ultimately taken on these bills in the 106th Congress.

In a recent report to Congress, the FCC decided to consider a notice of proposed rulemaking (NPRM) to be developed by its Wireless Telecommunications Bureau that will take "into consideration existing competitive conditions and technological developments that could affect the continued need for a cap."[61] On January 23, 2001, the FCC adopted an NPRM to reexamine the need for spectrum caps. The FCC is seeking comment on whether the spectrum caps for commercial mobile radio services should be eliminated, modified, or retained. Meanwhile, the Third-Generation Wireless Internet Act (S. 696), introduced by Senator Brownback on April 4, 2001, would require the FCC to lift spectrum aggregation limits (i.e., spectrum caps).

---

[61] FCC 00-456, CC Docket No. 00-175, Report, In the Matter of the 2000 Biennial Regulatory Review, released January 17, 2001.

SPECTRUM SCARCITY IN PRIVATE LAND MOBILE RADIO SERVICES

A set of non-federal radio communications systems called Private Land Mobile Radio Services (PLMRS), including public safety, special emergency, industrial, land transportation, and radiolocation services, are treated differently from spectrum licensees whose main activity is providing commercial wireless communications services to the public.[62] Prior to the 1990s, spectrum allocated for PLMRS was divided into frequency bands representing 20 different types of communications service areas (see Table 2). For each PLMRS service area, the FCC designated a "frequency coordinator," to manage the PLMRS licenses in its service area. The FCC granted applications for new PLMRS licenses based on recommendations from the frequency coordinators in a given PLMRS service area. As a result of an increased number of requests from the wireless telecommunications industry for new PLMRS licenses, and the realization that insufficient spectrum was available to meet future needs, the FCC in 1992 began a proceeding to revise the policies governing PLMRS licenses. This proceeding is known as PLMRS "re-farming" (a reference to the redistribution of frequency allocations and rules for their use).

Table 2. Public Land Mobile Radio Service Areas/Frequency Pools

Public Safety Pool	Industrial/Business Pool
Local Government Radio Service	Power Radio Service
Police Radio Service	Petroleum Radio Service
Fire Radio Service	Forest Products Radio Service
Highway Maintenance Radio Service	Film and Video Production Radio Service
Forestry-Conservation Radio Service	Relay Press Radio Service
Emergency Medical Radio Service	Special Industrial Radio Service
Special Emergency Radio Service	Business Radio Service
	Manufacturers Radio Service
	Telephone Maintenance Radio Service
	Motor Carrier Radio Service
	Railroad Radio Service
	Taxicab Radio Service
	Automobile Emergency Radio Service

---

[62] PLMRS licenses are governed under 47 CFR, Part 90.

In 1995, as a first step, the FCC adopted a plan for PLMRS licenses below 800 MHZ in specific frequency bands and geographic areas.[63] These bands were previously divided into smaller frequency channels that were separated by unused spectrum. The 1995 ruling utilized the unused spectrum by creating new channels between the existing channels, and allowed licenses to be granted on the new channels for the use of new narrowband communications devices. The FCC determined that by using a new digital radio technology and by placing limits on the use of the new licenses, the amount of interference introduced on existing channels by the new licensees would be negligible.

Then in 1997, another FCC ruling consolidated the 20 PLMR services into two broad service pools – one designated as public safety and the other as industrial/business (see Table 2).[64] The frequency coordinators in the public safety pool continued to manage the same frequency bands as prior to consolidation.[65] The frequency coordinators in the industrial/business pool, however, were now able to accept applications from any service area within the pool if they determine that spectrum is available and would not cause harmful interference to incumbent licensees. An exception to that provision was made for railroad, power, and petroleum companies, which were deemed to provide critical public safety-related communications. Anyone else who wants a PLMRS license for the frequencies previously allocated exclusively to those three service groups must seek the recommendation of the frequency coordinators responsible for those services. New PLMRS licenses for frequencies that were previously shared among PLMRS licensees in the industrial/business pool, however, could be assigned by any of the frequency coordinators, and would be shared with existing licensees. The ruling contained several other provisions for the management of PLMRS. One important provision allowed "centralized trunking systems," which resell private wireless communications services using PLMRS

---

[63] The specific bands are: 150-174 MHz (nationwide), 421-430 MHz (only in Detroit, Buffalo, and Cleveland), 450-470 MHz (nationwide), and 470-512 MHz (shared with UHF-TV, available in 11 U.S. cities). FCC 95-255, PR Docket 92-235, Report and Order and Further Notice of Proposed Rule Making, "In the Matter of Replacement of Part 90 by Part 88 to Revise the PLMRS and Modify Policies Governing Them;" released June 23, 1995.

[64] FCC 97-61, Second Report and Order, released March 12, 1997, (effective October 1997). The ruling applied only to PLMRS licensees in spectrum bands below 800 MHz, including the 150-174, 421-430, 450-470, and 470-512 MHz bands.

[65] An exception to this provision was made to allow any frequency coordinator in the public safety pool to use frequencies allocated to Local Government Radio Service, which typically overlaps services provided by police, fire, and other public safety services.

spectrum on a leased basis, to operate the PLMRS licensees below 80 MHZ, with certain limitations to protect users sharing the same spectrum.[66]

In June 1998, two of the frequency coordinators submitted to the FCC an "Emergency Request for Limited Licensing Freeze." The request was submitted by the United Telecommunications Council (UTC, representing the power industry) and the American Petroleum Institute (API, representing the petroleum industry) to prevent new PLMRS licensees in the business/industrial pool to be granted on certain frequencies that were previously shared between the power and petroleum industries and other business/ industrial pool services. UTC and API argued that since the 1997 ruling went into effect, their industries had experienced an increased number of incidents of interference from other business/industrial pool licensees which posed a dangerous threat to power and petroleum operations. This request was followed by a Petition for Rulemaking, submitted by UTC, API, and the Association of American Railroads (AAR, represent the railroad industry). The petition asked the FCC to establish a new third pool for PLMRS licenses involved in public safety-related services, to protect those services from interference and encroachment by new industrial and commercial communications systems.

In April 1999, the FCC revised its 1997 ruling, requiring all PLMRS frequencies (both shared and exclusive) assigned to the power, petroleum, and railroad services prior to the adoption of the 1997 ruling to be coordinated by the frequency coordinators previously responsible for these services.[67] Applicants for these frequencies must obtain the concurrence of UTC, API, or AAR, as appropriate. The ruling also provided a similar protection to frequencies previously allocated to the former Automobile Emergency Radio Service, whose frequency coordinator is the Automobile Association of America.

In July 1999, two other frequency coordinators, the Manufacturers Radio Frequency Advisory Committee (MRFAC, representing the manufacturing industry) and Forest Industries Telecommunications (FIT, representing the forest products industry), petitioned the FCC to reconsider aspects of the 1999 ruling. The petition argued that the ruling gave a special status to frequencies allocated on a shared basis with the power, petroleum, and railroad radio services to the detriment of others in the industrial/business pool. The petition requested that the FCC stay the

---

[66] A trunking system has the ability to automatically search all available radio frequencies for one that is not being used.

[67] FCC 99-68, Docket 92-235, Second Memorandum Opinion and Order, released April 13, 1999.

effective date of the ruling until this issue was addressed. Some argued that if auctions are implemented for some future PLMRS licenses, a radio service that has a higher public safety status could possibly be exempted from an auction. Thus, creating a special pool for public safety-related services could create economic advantages for some licensees. In August 1999, the FCC granted a stay of its rules as requested by MRFAC and FIT until a final ruling on the matter was made.[68]

Legislation was introduced in the 106th Congress addressing the PLMRS issue (H.R. 866, introduced February 25, 1999) on which no action was taken. The bill would have created an advantage for the power or petroleum radio services over other Industrial/Business pool licensees by discontinuing licensing on frequencies formerly allocated to or near the power or petroleum radio services.

In November 2000, as part of its proceeding examining its auction authority, the FCC decided not to create a separate PLMRS license pool.[69] The FCC's Order also modified the PLMRS rules to allow licensees in the 800 MHZ Business and Industrial pool and the Land Transportation service area to convert their spectrum to commercial use under certain restrictions. The ruling also establishes that some PLMRS licenses (along with all private radio services) could be subject to auctions in the future if certain conditions (such as mutual exclusivity) apply. Although the FCC's ruling might have resolved the issues of managing PLMRS licenses, demand for PLMRS spectrum continues to be high. Disputes among the industries could again come to the attention of Congress if the parties are not satisfied with the ruling.

ATTEMPTS TO USE SPECTRUM AUCTION REVENUE FOR OTHER PROGRAMS

Current FCC policy, in accordance with current statue, is that potential revenue should not be used as the main argument for auctioning spectrum. Although many agree with that policy, revenue from spectrum auctions was used in the Balanced Budget Act of 1997 to help reduce the federal deficit. In addition, many bills were introduced in the 105th Congress to used

---

[68] FCC 99-203, PR Docket 92-235, Fourth Memorandum Opinion and Order, released August 5, 1999.

[69] FCC 00-403, WT Docket No. 99-87, Report and Order and Further NPRM, released November 20, 2000.

spectrum auctions to offset various spending programs. None of those bills was enacted, and such attempts declined in the 106th Congress. This could be a result of decreased fiscal pressures due to the budget surpluses experienced during the past several years, as well as the acknowledgment that recent auctions have produced less revenue than the initial auctions.

However, at least one bill in the 106th Congress (S. 2762, introduced June 21, 2000) did contain a Sense of Congress statement advocating the use of spectrum auction revenue for a specific purpose. That bill, on which no action was taken, stated that "resources available through the auction of the analog [television] spectrum should be tapped to fund the development of a new educational and cultural infrastructure that utilizes today's technologies..." It is possible that legislation could be introduced again in the 107th Congress to use auction revenue for specific purposes other than reducing the federal debt, especially if federal budget pressures are exacerbated.

Chapter 5

OTHER SPECTRUM-RELATED LEGISLATION IN THE 106TH CONGRESS

In addition to the legislation discussed in this report, several other bills were introduced in the 106th Congress that contained provisions concerning spectrum management. Some of them became law, while many did not.

- S. 376, introduced February 4, 1999 (the companion bill, H.R. 3261, was introduced in the House November 9, 1999) contained a provision prohibiting the FCC from assigning orbital locations or spectrum licenses to international or global communications services through the use of auctions. It further directed the President to oppose in the International Telecommunications Union and other international fora the use of auctions for such purposes. Enacted as the Open-market Reorganization for the Betterment of International Telecommunications Act (**P.L. 106-180**) on March 17, 2000.

- H.R. 783 (introduced February 23, 1999) sought to ensure the availability of spectrum to amateur radio operators by prohibiting reallocations of amateur radio service and amateur satellite service spectrum unless the FCC provides equivalent replacement spectrum. Referred to House Commerce Committee, Subcommittee on Telecommunications, Trade and Consumer Protection. Companion bill S. 2183 (introduced March 6, 2000) referred to Senate Committee on Commerce, Science and Transportation. No further action.

- H.R. 879 (introduced February 24, 1999) would have exempted licenses in the instructional television fixed service from auctions. Referred to House Committee on Commerce, Subcommittee on Telecommunications, Trade and Consumer Protection. No further action.

- H.R. 1554 (introduced April 26, 1999) contained a provision (Sec. 5008, "Community Broadcasters Protection Act") and companion bill S. 1547 (introduced August 5, 1999) seeks to preserve low-power television (LPTV) stations that provide community broadcasting by directing the FCC to grant qualifying LPTV stations special "class A" licenses and find alternative spectrum on which to locate those stations wherever full power stations are given priority to their spectrum in the transition to DTV. Incorporated into the Intellectual Property and Communications Omnibus Reform Act of 1999 (S. 1948, Section 5008, which was incorporated by cross-reference in the conference report H.Rept. 106-479 to H.R. 3194, FY2000 Consolidated Appropriations bill) and enacted (**P.L. 106-113**) on November 29, 1999.

- H.R. 2379 (introduced June 29, 1999) sought to ensure that biomedical telemetry operations are provided with adequate spectrum to support that industry's existing and future needs, and requires a report from the FCC on future spectrum needs of telemedicine and telehealth providers. Referred to House Committee on Commerce, Subcommittee on Telecommunications, Trade and Consumer Protection. No further action.

- H.R. 2630 (introduced July 29, 1999) NTIA Reauthorization Act, contained a provision directing NTIA to conduct an assessment of spectrum reallocations to non-federal use and the implications of such reallocations for affected federal agencies. Although the bill was not enacted, a similar provision was added to the FY2000 Defense Authorization Act (**P.L. 106-65**).

- S. 1824 (introduced October 28, 1999) would have directed the FCC to: (1) identify and allocate at least 12 MHZ of spectrum located between 150-200 MHZ for use by private wireless licensees on a shared-use basis; (2) reserve at least 50 percent of the reallocated spectrum for private wireless systems; and (3) reallocate and assign licenses for such spectrum. It further required the FCC to devise a schedule for payments to the Treasury for shared spectrum used by private wireless systems, and adopt a payment schedule in the public interest. Referred to House Committee on Commerce. No further action.

- H.R. 3615 (introduced March 10, 2000) would have provided loan guarantees to rural television stations for improvements related to producing local broadcasts in underserved areas, but prohibited the use of the funds for spectrum auctions. Reported by the House Committee on Agriculture, Commerce, and Rules. Passed the House (amended) April 13, 2000. No further action.

- S. 2454 (introduced April 13, 2000) authorizes low power television (LPTV) stations to provide "data-casting" (e.g., financial or Internet-related) services to subscribers. Referred to Senate Committee on Commerce. A similar provision was inserted in the FY2001 Omnibus Consolidated and Emergency Supplemental Appropriations Act (H.R. 4577), enacted December 21, 2000 (**P.L. 106-554**).

- H.R. 4758 (introduced June 26, 2000) would have prohibited the FCC from applying limits on spectrum aggregation to any license for commercial mobile radio service granted in an auction held after January 1, 2000, or to any subsequent application for the transfer or assignment of such a license. Referred to House Committee on Commerce, Subcommittee on Telecommunications, Trade and Consumer Protection. No further action.

Chapter 6

Spectrum-Related Legislation in the 107th Congress

H.R. 817 (Bilrakis)

Amateur Radio Spectrum Protection Act of 2001. Prevents the FCC from reallocating or diminishing frequencies used by amateur radio and amateur satellite services. Introduced March 1, 2001; referred to Committee on Energy and Commerce.

S. 549 (Crapo)

Amateur Radio Spectrum Protection Act of 2001. Prevents the FCC from reallocating or diminishing frequencies used by amateur radio and amateur satellite services. Introduced March 15, 2001; referred to Committee on Commerce, Science, and Transportation.

S. 696 (Brownback)

Third-Generation Wireless Internet Act. Prohibits the Federal Communications Commission from applying spectrum aggregation limits to spectrum assigned by auction after December 31, 2000. Introduced April 4, 2001; referred to Committee on Commerce, Science, and Transportation.

Chapter 7

CONCLUDING REMARKS

The growth in demand for wireless services has been unprecedented, with estimates that, by 2002, wireless users will number up to one billion worldwide.[70] Whether the U.S. wireless industry will keep up with or surpass the growth and penetration rates of other industrialized nations, depends in part on how the radio spectrum is managed both here and abroad. The wireless services industry is highly dynamic, spurred by the growth of electronic commerce and development of wireless Internet access services. Federal, state, and local government spectrum needs are also increasing as those services strive to maintain the same level of technology as the private sector.

The pace of the wireless revolution demands that new spectrum resources be found quickly. The challenge is to manage the spectrum resources to maximize the efficiency and effectiveness of meeting these demands without disenfranchising incumbent spectrum users, while ensuring that competition is maximized and that consumer, industry, and government groups are treated fairly. Congress will likely play a significant role in ensuring that those challenges are met.

---

[70] Testimony of Tom Sugrue, Chief, Wireless Telecommunications Bureau, FCC, before the House Commerce Committee, Subcommittee on Telecommunications, Trade and Consumer Protection, July 19, 2000.

APPENDIX 1. SPECTRUM LICENSES AUCTIONED TO DATE

License type and geographic area	Amount of spectrum per license and its uses	Number of licenses sold	Date closed	Net High bids, $millions
Narrowband: nationwide	50 kHz (paired), 50/12.5 kHz, and 50 kHz: paging, messaging	10	7/94	617.0
IVDS: MSAs	0.5 MHZ: interactive data	594	7/94	213.9
Narrowband PCS: regional	50 kHz (paired), 50/12.5 kHz: paging messaging	30	11/94	392.7
Broadband PCS A&B blocks: MTAs	30 MHZ: mobile voice and data	99	3/95	7,019.4
DBS: nationwide (two separate auctions)	Shared spectrum* Subscription television service	2	1/96	734.8
MDS: BTAs	Heavily encumbered spectrum* Subscription TV broadcast	493	3/96	216.2
900 MHZ SMR: MTAs	25 kHz: mobile dispatching	1,020	4/96	204.3
Broadband C-block PCS: BTAs	30 MHZ: mobile voice and data	493	5/96	9,197.5
Broadband C-block PCS re-auction: BTAs	30 MHZ: mobile voice and data	18	7/96	904.6
Broadband PCS blocks D, E, and F: BTAs	10 MHZ: mobile voice and data	1,472	1/97	2,517.4
Unserved cellular areas: MSAs, RSAs	25 MHZ (encumbered): mobile voice and data	14	1/97	1.8
Wireless communications service: Major/Regional Economic Areas	10 MHZ and 5 MHZ: multiple wireless uses	126	4/97	13.6
Digital audio radio services: nationwide	12.5 MHZ: satellite radio broadcasting	2	4/97	173.2
Upper 800 MHZ SMR: Economic Areas	1 MHZ, 3 MHZ, and 6 MHZ: mobile voice and data	524	12/97	96.2
LMDS: BTAs	1,150 MHZ and 150 MHZ: fixed voice, data, and video	864	3/98	578.7
220 MHZ band: economic area/grouping, nationwide (two separate auctions)	100 kHz and 150 kHz (encumbered): voice, data, paging, fixed services	915	10/98 6/99	23.6
VHF Public Coast: Public Coast station areas	25 kHz (paired): maritime/ship fixed/mobile communications	26	12/98	7.5
C, D, E, and F Block Broadband PCS: BTAs	30 MHZ, 15 MHZ, and 5 MHZ (paired): mobile telephone service	302	4/99	412.8

License type and geographic area	Amount of spectrum per license and its uses	Number of licenses sold	Date closed	Net High bids, $millions
Locations and Monitoring Service: Economic Areas	6 MHZ, 2.25 MHZ, and 5.75 MHZ with 250 kHz (paired): radio signals to determine location/use of mobile phones	289	3/99	3.4
LMDS re-auction: BTAs	1,150 MHZ and 150 MHZ: multiple wireless uses	161	5/99	45.1
"Closed" broadcast station construction permits: no geographic limit defined	10 kHz (AM), 200 kHz (FM), 6 MHZ (television, LPTV): radio and TV broadcasting	115	10/99	57.8
929-931 MHZ paging services: major economic areas	20 kHz: paging/messaging services, data transmission	985	3/00	4.1
Broadcast construction permits (three separate auctions)	200 kHz (FM), 6 MHZ (television, LPTV): radio and TV broadcasting	4	10/99 3/00 7/00	20.2
39 GHz band licenses: economic areas	50 MHZ (paired): fixed/mobile two-way communications	2,173	5/00	410.6
700 MHZ guard band: major economic areas	2 MHZ (paired) and 1 MHZ (paired): leased wireless services subject to strict interference limits	96	9/00	519.9
800 MHZ SMR (general category): economic areas	1.25 MHZ (split into six channels): voice/data/paging and other wireless services	1,030	9/00	319.5
Reclaimed C&F block PCS licenses, 1890-1975 MHZ	Two 15 MHZ (paired) or three 10 MHZ (paired) (C-block, 10 MHZ (paired) F-block; mobile voice and data	2,800	12/00	29.0
700 MHZ guard band: unsold licenses from previous 700 MHZ guard band auctions	2 MHZ (paired) and 1 MHZ (paired): leased wireless services subject to strict interference limits	422	1/01	16,857.0
VHF Public Coast (156-162 MHZ) and LMF licenses (902-928 MHZ)	6 MHZ Public Coast (marine communications); 6, 2.25, 5.75 MHZ LMF (radiolocation)	8	2/01	21.0
Total				**$41,614.1 M**

Source: CRS, based on data from the FCC

* Due to the DBS rules for spectrum channelization and the existence of many licensees encumbering the MDS spectrum, the amount of spectrum for the DBS and MDS auctions cannot be easily established.

APPENDIX 2. ONGOING AND FUTURE SPECTRUM AUCTIONS

License Type: Geographic Area	Frequency Range	Auction Date
700 MHZ Band	746-764 MHZ and 776-794 MHZ	September 12, 2001
Lower 700 MHZ band	698-746 MHZ	Not yet scheduled
FM Broadcast	88-108 MHZ	December 5, 2001
AM Filing Window	0.535-1.065 MHZ	Not yet scheduled
24 GHz (DEMS)	Around 24 GHz	Not yet scheduled
4.9 GHz	4940-4990 MHZ	Not yet scheduled
Narrowband PCS: MTA and BTA	50 KHz paired licenses in the 900 MHZ range	Not yet scheduled
LPTV	512-806 MHZ	Not yet scheduled
Paging (lower channels	Licenses of varying sizes between 35-930 MHZ	June 26, 2001
218-219 MHZ (formerly IVDS): MSA and RSA	500 kHz licenses at 218 and 219 MHZ	Not yet scheduled
Public Coast and Location and Monitoring Services	156-162 MHZ and 904-928 MHZ	June 6, 2001
GWCS	4600-4685 MHZ	Not yet scheduled

Source: CRS, based on data from the FCC.

INDEX